Everyday Engineering

Inside Technology
edited by Wiebe E. Bijker, W. Bernard Carlson, and Trevor Pinch

Janet Abbate, *Inventing the Internet*

Charles Bazerman, *The Languages of Edison's Light*

Marc Berg, *Rationalizing Medical Work: Decision-Support Techniques and Medical Practices*

Wiebe E. Bijker, *Of Bicycles, Bakelites, and Bulbs: Toward a Theory of Sociotechnical Change*

Wiebe E. Bijker and John Law, editors, *Shaping Technology/Building Society: Studies in Sociotechnical Change*

Stuart S. Blume, *Insight and Industry: On the Dynamics of Technological Change in Medicine*

Geoffrey C. Bowker, *Science on the Run: Information Management and Industrial Geophysics at Schlumberger, 1920–1940*

Geoffrey C. Bowker and Susan Leigh Star, *Sorting Things Out: Classification and Its Consequences*

Louis L. Bucciarelli, *Designing Engineers*

H. M. Collins, *Artificial Experts: Social Knowledge and Intelligent Machines*

Paul N. Edwards, *The Closed World: Computers and the Politics of Discourse in Cold War America*

Herbert Gottweis, *Governing Molecules: The Discursive Politics of Genetic Engineering in Europe and the United States*

Gabrielle Hecht, *The Radiance of France: Nuclear Power and National Identity after World War II*

Kathryn Henderson, *On Line and On Paper: Visual Representations, Visual Culture, and Computer Graphics in Design Engineering*

Eda Kranakis, *Constructing a Bridge: An Exploration of Engineering Culture, Design, and Research in Nineteenth-Century France and America*

Pamela E. Mack, *Viewing the Earth: The Social Construction of the Landsat Satellite System*

Donald MacKenzie, *Inventing Accuracy: A Historical Sociology of Nuclear Missile Guidance*

Donald MacKenzie, *Knowing Machines: Essays on Technical Change*

Donald MacKenzie, *Mechanizing Proof: Computing, Risk, and Trust*

Maggie Mort, *Building the Trident Network: A Study of the Enrollment of People, Knowledge, and Machines*

Paul Rosen, *Framing Production: Technology, Culture, and Change in the British Bicycle Industry*

Susanne K. Schmidt and Raymund Werle, *Coordinating Technology: Studies in the International Standardization of Telecommunications*

Dominique Vinck, editor, *Everyday Engineering: An Ethnography of Design and Innovation*

Everyday Engineering
An Ethnography of Design and Innovation

edited by Dominique Vinck

with contributions by
Éric Blanco
Michel Bovy
Pascal Laureillard
Olivier Lavoisy
Stéphane Mer
Nathalie Ravaille
Thomas Reverdy

The MIT Press
Cambridge, Massachusetts
London, England

First MIT Press paperback edition, 2009.

This edition ©2003 Massachusetts Institute of Technology.

This work originally appeared under the title *Ingénieurs au quotidien: Ethnographie de l'activité de conception et d'innovation* (©1999 Les Presses Universitaires de Grenoble).

Set in New Baskerville by The MIT Press.

Library of Congress Cataloging-in-Publication Data

[Ingenieurs au quotidien. eng]
Everyday engineering : an ethnography of design and innovation / edited by
Dominique Vinck ; with contributions by Éric Blanco . . . [et al.]
p. cm. — (Inside technology)
Includes bibliographical references and index
ISBN 978-0-262-22065-1 (hc : alk. paper), 978-0-262-51264-0 (pb)
1. Engineering—Social aspects. I. Vinck, Dominique. II. Blanco, Éric.
III. Series.

TA157 .I475   2003
620—dc21

20002070157

Contents

6
A Prototype Culture: Designing a Paint Atomizer   119

III
Technical Writing Practices

7
Writing Procedures: The Role of Quality Assurance Formats   137

8
The Role of Graphical Representations in Inter-Professional Cooperation   159

9
Rough Drafts: Revealing and Mediating Design   177

Everyday Engineering

Introduction

We live in an era of paradoxes. On the one hand, we are faced with an ever-growing array of technological developments that affect communication, work, travel, domestic and leisure activities, and political and ethical debates. We cannot envisage life without them. They structure the world we live in. On the other hand, in many ways we do not understand these technologies. The surprising thing is that this is true for technological professionals as well as for laypeople. Our situation is characterized both by our ignorance of new technologies and by our faith in them. Sometimes technology is brushed aside as something belonging to another world. At other times, when our attention is focused on humans, technology may appear inhuman or superhuman. It may seem to belong to an area reserved for specialists. With its principles, laws, methods, and models, technology can seem cold, rational, boring, and inaccessible. It is taught and presented in this way to specialists. It may be packed into a set of rational theories, but this does not mean that it is understood any better.

In this book we propose another approach to our technical reality. We invite the reader to open the doors of plants, design offices, and laboratories so as to see how things really are done. A different vision of technology will emerge—a vision that technicians should find easy to understand because it will be based on their day-to-day life. We will see, for instance, engineers busy tinkering with high-tech prototypes, struggling with unsuitable software programs, and even negotiating the installation of a new waste container with stubborn salespeople.

While providing testimony and socio-technical analysis, we also give engineers some pointers concerning technical processes and associated tools. We do not aim to provide a single answer or a guide to the best practices; rather, we hope to bring to light some new facts about the complexity of the actual situations and practices an engineer must face. We try to show that these situations and practices can be approached through ethnographic methods that generate know-how and instruments for action.

The Need to "Go out into the Field"

Our hypothesis is that, in order to understand modern technology, appreciate what engineers do, and fully grasp the scope of the industrial changes that are taking place, it is necessary to go out into the field to observe and analyze current practices. We demonstrate that "going out into the field" is also of great importance to novice engineers. This is why we have chosen to look at many technical practices in the field, to study the tools and the various ways they are used, and to take into account the real means for action.

Despite the fact that technology involves many people, only a few are actually technicians. Technology covers a multitude of objects, many of which are very ordinary (modeling clay, paper and pencil, production rejects). It involves symbols that are often very simple (lines, circles, figures), though they are sometimes designated by strange names. Technology implies organizations that are somewhat complicated. It encompasses passions, habits, and values that engineers are ready to fight for, economic constraints imposed from above, and social projects (such as building a more ecologically sound society).

We invite the reader to enter various worlds of technology. The authors, most of them engineers, have already done so. Although they have taken modeling clay in their hands, they have distanced themselves from what they have seen or experienced. They have observed, but it has been *participative* observation. We use their reports and analyses, and consequently we offer their views of the various practices they came across. Each technology is looked at by analyzing small details, gestures, corridor discussions, rough drafts that have been changed several times, and the individuals who manage to produce results (or at least something that works).

Understanding Technical Action

We propose an ethnographic approach to technologies that takes objects into account just as much as human beings. We show that this approach makes it possible to analyze and theorize ordinary technical practices in a most productive way.

The objective is to understand. This is why we wanted to look at things in detail, to "soak it all up," to describe these things and apply our ethnography to them, to be surprised by them, to question them, and then take a step back through discussion, reading, and writing. What follows is not a series of flat, neutral descriptions; rather, it is a set of attempts to interpret situations and to compare the various points of view we encountered against observations from our field studies.

The aim is to understand technical action. This form of action cannot be reduced to neutral principles and objectives concerned only with scientific and technological knowledge. It is therefore of interest to sociologists and anthropologists alike. The goal of anthropology is to understand the human being. It looks at social practices because they provide information about what makes sense to people. This sense is very much linked to what they do, to the actions they carry out, and to the results and performances they obtain. This means that we must study what happens during action as well as what happens as a result of action, both in the order in which they happen and in the sensible order, given that the two things are practically indissociable for human beings. Another of our hypotheses is that it is not possible to understand human beings and society without taking the effectiveness of their action into account. This technical effectiveness is productive in terms of both meaning and identity, whereas ineffectiveness results in weariness of the subject in question. Ethnography of technology does not just boil down to a cold description of objects and mechanisms void of human presence. On the contrary, technical action is entirely composed of meaning and performance.

An Action-Focused Understanding

We were often actively involved in the field, not only trying to glean information in order to note its scientific value and thus gain knowledge but also accompanying the actors in their projects—projects that are, in fact, our own: improving design methods and taking into account issues that have until now been neglected by companies (environmental issues, for example). Taking ethnographic and sociological approaches, we set out to encourage the actors we studied, particularly the engineers, to look at their own practices differently.

Here are some of the questions we asked ourselves:

- What makes something work, and why does it produce the effects it does?

- How can we understand what engineers do, and particularly their design activities, in order to suggest new methods and tools?

- What kind of new instrument designs can be proposed, not according to general principles or fantastic technical solutions, but on the basis of effective use and practices?

These questions were also asked by the actors in the field. There are no simple answers to such questions. It is not tenable to say that something works thanks to such and such a technical principle, type of organization, or power relationship, or simply because society was ready to receive such a new idea. We have also decided not to put forth prescriptive models in response to our observations. Because the specific contexts of the situations observed provide too complex a picture of the activities, the best way to deal with these questions is to go into more detail and analyze the different types of mediators.

This work is addressed primarily to future engineers, the individuals on whom society relies to "make things work." Going beyond their training in the principles of technical rationality, we aim to give them a concrete understanding of local, in-the-field action. In short, our purpose is to give them an idea of some of the forces that are behind effective action.

Understanding Oriented toward the Production of New Tools

It is in the nature of sociology to question the very structure of engineering tools, especially the tools that aid in design work. Owing to their double role, most of the present authors take a view of these tools that is centered on use and action. This leads them to suggest certain types of tools in preference to conventional product-simulation tools, which focus on modeling and physics and which do not offer actors a means of coordinating and creating cohesion in the design process. Seen as a model or a technical representation, the tool is also perceived here as a social mediator and as an instrument of organization dynamics.

Through the various situations with which they have been confronted, the authors have built CAD systems and modified the uses made of them, supplied tool specifications, and suggested new procedures. This in-the-field action has contributed to producing original and hybrid results, moving from specific ideas to more general ones. Thus, our thinking about new types of design aid tools, and more generally about the production of technical tools, springs from experience in the field.

An Inter-Disciplinary Production

This book is the result of collective research into design and technological implementation processes. The authors are associated with laboratories that have dared to embark on inter-disciplinary activities; that is, they have decided to look at the same subject or the same field from different points of view.[1] This has meant that, since the beginning of the 1990s, sociologists and mechanical engineers from Grenoble have learned to work together, with the help of Serge Tichkiewitch (a professor of mechanics at the École Nationale Supérieur d'Hydraulique et de Mécanique de Grenoble) and Alain Jeantet and Henri Tiger (sociologists and Centre National de la Recherche Scientifique researchers at the Center de Recherche: Innovation Socio-Technique et Organization industrielle), through joint research seminars, association in various research contracts, and co-management of DEA and Ph.D. dissertations. All these activities are managed within the framework of a recognized institute: the Institut de la Production et des Organisations industrielles, a Grenoble-based academic body whose purpose is to promote interdisciplinary research by financing research and seminars.

Most of the authors have backgrounds in both social sciences and engineering. They were supervised or assisted in their research by both mechanical engineers and sociologists. This means that they are well aware of the problems, concepts, and methods that were available to guide them. They have also invented, according to their own areas of interest, several ethnographic positions; that is, there was flexibility in the type and extent of insertion and participation and the type of report produced. The engineers among the authors agreed to make certain on-site detours with the sociologists to incorporate ethnographic concepts and methods. They learned how to take notes and keep a diary, how to debate with sociologists, and how to write detailed reports. Without this help, their experiences might never have been shared, and their observations might have led to nothing more than a handful of recommendations, models, or general principles. Technicians who have learned the discipline necessary for ethnographic observation and writing are probably best placed to comment on the advantages of such an approach.

All the authors are affiliated with the University of Grenoble.

Éric Blanco is a lecturer in mechanics. He holds a doctorate in industrial engineering with a major in mechanics. His dissertation was supervised by Olivier Garro (mechanics) and Alain Jeantet (sociology). Attached to the Sol-Solide-Structure laboratory, he works on the use of criteria in design processes.

Michel Bovy holds a doctorate in environmental public management. His dissertation was supervised by Marc Mormont and Dominique Vinck (sociology). Attached to the Socio-Economie Environnement et Développement laboratory of the Fondation Universitaire Luxembourgeoise, he works on the implementation of the Citizen's Public Environment and Participation Policy.

Pascal Laureillard is a mechanical engineer. He holds a doctorate in Industrial Engineering with a major in mechanics. His dissertation was supervised by Jean-François Boujut (mechanics) and Alain Jeantet (sociology). Attached to the Sol-Solide-Structure laboratory, he works on inter-trade coordination in design processes.

Olivier Lavoisy holds a doctorate in industrial engineering with a major in industrial engineering and sociology. His dissertation was supervised by Dominique Vinck and Alain Jeantet (sociology) and Olivier Garro (mechanics). Attached to CRISTO, he works on the history of sociology in industrial design.

Stéphane Mer is a mechanical engineer with a doctorate in industrial engineering and a major in mechanics. His dissertation was supervised by Serge Tichkiewitch (mechanics) and Alain Jeantet (sociology). He works on the characterization of the social worlds of design.

Thomas Reverdy is a lecturer in sociology. Trained as an industrial engineer, he holds a doctorate in industrial engineering with a major in economy and sociology. His dissertation was supervised by Denis Segrestin and Dominique Vinck (sociology). A member of CRISTO, he works on the implementation of management referential in industrial environmental management.

Nathalie Ravaille is an engineer who did postdoctoral work on Company Research and Development. Attached to the Sol-Solide-Structure laboratory, she works on improving design methods.

Dominique Vinck is a professor of sociology at the Université Pierre Mendès-France and the Institut National Polytechnique de Grenoble. Trained as a chemical and agricultural engineer at the University of Gembloux in Belgium, he holds a degree in philosophy and a doctorate in innovation socio-economics. As a member of CRISTO, he works on the analysis of design processes and the implementation of quality approaches in research and care services.

How to Use the Book

This book looks at the everyday work of engineers and technicians. Particular attention is paid to design, change management, and innovation.

A great deal has already been written about innovation, design, and change. In spite of this, precious little is actually known about design-related practices. The literature on these subjects is far more prescriptive than descriptive. There is certainly no lack of management and method-ology studies with examples of good practices and approaches to imple-ment. Such studies often present models that correspond to ideal objectives, but what they say about how to proceed hardly tells us any-thing about how things actually work. In this book we take a detour in order to observe, analyze, and describe effective practices. We are sure that this detour is not only useful but necessary for those setting out to design new means of action.

We have tried to write in such a way that the chapters can be read inde-pendently. Of course, there is much to be derived from reading the book in its entirety.

For selective reading or classroom use, the book offers a series of detailed reports describing the various practices observed. These case studies can be used to train engineers or students of the social aspects of science and technology. Readers can find the objects, fields, and themes that interest them in the table of contents. Some chapters (4, 5, 6, 8) deal with mechanical design in manufacturing industries, others with instru-ment design and production of scientific knowledge in research (chap-ter 1) or software debugging or validation (chapter 2). Some chapters look at innovative projects within companies (for example, chapter 7 examines the implementation of environmental management); others go even further (for example, chapter 3 addresses the implementation of a system for separating household waste).

In a book of this kind, the complexity of technical practices must be assessed at the start. Manuals on design methodology or management tend to present simplified models. The teaching of science is based on a desire to summarize knowledge in the form of laws and general princi-ples. These models and laws stem from an analytical and reductionist approach, and such an approach is necessary for mastering industrial phenomena and action. However, young engineers often are given an oversimplified and irrelevant representation of industrial activity. We must never forget that practices differ considerably. Real work situations are always complex. Once again, it is not enough just to state this com-plexity; we have to show what it actually consists of.

Part I introduces the socio-technical complexity of technical practices; it should help young engineers to assess the nature of this complexity and to better situate their contributions and the benefits of their tools.

In chapter 1, we follow a young mechanical engineer assigned to a design office and entrusted with designing a simple technical element:

a wall to separate two devices. We gradually discover the technical and socio-technical complexity this young engineer must take into account to succeed in designing the wall. This complexity will lead him to consider a growing number of objects and actors. We will see through him that the object is one node in a network. The more this engineer enters into the technical content of the object, the wider the range of people he has to consult. Conversely, the more he discovers about the socio-technical network surrounding the object, the more he has to analyze it from a different point of view.

In chapter 2, we continue the investigation into complexity by identifying the stakes in the tool-design process. We see that a tool (in this case a functional dimensioning software package) plays a much more important role than is suggested by its specifications. It lies at the heart of unforeseen industrial stakes. A better understanding of these stakes, therefore, is of help in evaluating and perhaps in redirecting the company's action. Discovering the stakes requires both entering into the technical content and broadening the observation horizon. It gradually becomes clear that a tool is sometimes different from what was initially imagined. The type of instrument is not given a priori; on the contrary, it emerges little by little in the course of design, validation, and use.

Chapter 3 tells the story of an innovative project having to do with the separation of household wastes. By showing the social complexity that emerges as the project progresses, it addresses the same type of socio-technical complexity just mentioned. This chapter pays particular attention to the role of the material object, even though a trash can may appear to be somewhat trivial. The chapter raises questions about the overlap between the social actors' game and the relatively active role of the object. It shows that this object reflects and conveys the actions and compromises of the numerous actors involved; it relates the actors' action while mediating it (that is, it translates the action). This chapter also shows that to conclude an analysis of the action based on the objects alone is not enough. Objects are regularly surpassed by actors.

The chapters in part II examine social worlds, cultures, and technical action practices. Simply saying "design office," "structural engineer," or "research laboratory" is not tantamount to understanding what they do. The corresponding social worlds must be analyzed in more detail, since very different realities can hide behind the labels.

Chapter 4 takes us into the design office world of the structural engineer. It outlines an approach to characterizing this specific universe of action and the human, material, and textual entities within it on the basis of three concepts: action-based logic, scale of values, and shared knowledge. It shows how the differences among the various entities identified cause controversies between actors, how such controversies turn a design office into a dynamic social world, and why it is appropriate to propose a new way of looking at relations among design actors.

Chapter 5 concentrates on what guides designers in their activities. In addition to principles, tools, and methods, the activity of design is linked to know-how and conventions. Largely implicit, this know-how and these conventions gradually structure the action's technical instrumentation. Thus, design is a part of a collective history. By comparing the work of designers in two companies in the same field, we can characterize the "cultural basis" that affects how problems are posed and solved.

Chapter 6 also reports on the designer's universe, although this time the universe is characterized by the importance given to developing and using prototypes. Particular attention is paid to the mediations set up by these prototypes in the overall socio-cognitive design process. As this chapter shows, an unfinished object can bring actors together very quickly. Despite its state of partial completion, a prototype provokes irreversible situations with respect to industrial strategy.

The chapters in part III concern writing practices, which would seem to have little impact but which in fact are particularly decisive in the achievement of technical performance. Engineers and technicians spend a lot of time writing, jotting things down, and drawing. We try to pinpoint the role of these writing practices and their end products.

Chapter 7 looks at writing forms developed within the context of quality assurance. It shows that writing, aside from being a record of what is done, is linked to rules, forms, organization, and implementation that influence its effectiveness. This chapter also deals with difficulties, risks, and dead ends encountered in writing.

Chapter 8 deals with technical graphics as a mediator of inter-trade exchanges. With the emergence of concurrent engineering, the cooperation between designers and between trades in terms of logic and constraints has become crucial. Organizational and technical innovations (project platforms, shared databases, and so on) have been set up to enhance such cooperation, but these are by no means sufficient to create real interactivity and understanding among players. This is why it is important to question the objects that actors use when they are trying to agree. This chapter presents a series of graphical forms and ways of cooperating over the course of projects. Showing that inter-trade cooperation requires re-thinking of certain graphical objects, this chapter underlines the notion of "handholds" for actors and explores the idea of entity cooperation.

Chapter 9 focuses on the enlightening and mediating role of rough drafts in the design process. Rough drafts give the observer access to part of the socio-cognitive process. They are intermediary representations produced and used by the actors. Where design action is concerned, they make it possible to renew the modeling process.

Each chapter concludes with an operational summary that helps the reader to focus on the main lessons to be learned from the observations and analyses presented in that chapter. The principal benefits to be had from this approach are anthropological (What can be learned about technical action?) and operational (What does this imply for the action?).

In the epilogue, which is specifically addressed to specialists in social studies of science and technology, we re-situate the authors' ethnographic writing approaches. The ethnography proposed in the epilogue favors an approach based on both technical and scriptural mediation. The questions raised here revolve around how objects in sociological analysis can be taken into account. The aim is to structure our thoughts on the sociological status of objects, and the ideas are presented as an opening to debate.

Some readers may notice our use of the personal pronoun 'he' and the possessive adjective 'his' in generic references. We do not intend to convey any gender-specific values or preferences, and in the design offices we studied the actors were indeed all men.

I

Ordinary Technical Tasks and Their Complexity

The chapters in this part focus on evaluating the complexity that is inherent in technical practice. A young engineer often has a simplistic or even an inappropriate view of industrial activity. Real work situations are complex. Rather than simply confirm this complexity, these three chapters aim to show what it is made of.

1

Socio-Technical Complexity: Redesigning a Shielding Wall

Dominique Vinck[1]

Before embarking on a placement[2] in a design and engineering office, the young engineer does not really understand the complexity of the work awaiting him. Of course, he is ready to do complicated operations that must be dealt with at a high level of abstraction. He has also been trained to handle fairly sophisticated models and tools. He knows that he is bound to run up against difficult technical problems. Nevertheless, he has a certain number of working methods and tools under his belt that will get him out of many a difficult situation. He has the capacity to analyze problems, break them down into essential parts, and then model them. This ability to simplify things is supposed to help him get through the most complicated challenges. At least this is what he has been taught.

Yet the young engineering student still has to learn exactly how complex ordinary technical work really is. A placement period lasting just a few months will prove to be a real eye opener. He may have thought that an engineer's work is mainly technical, but he will quickly realize that, in reality, things are much more complex than that. He will also find that, if he wants to be an efficient engineer and get technically satisfactory results, he will have to decode and take into account only what appears to be real.

The aim of this chapter is thus to map and document the changing vision of young engineers after their entry into the industrial world. To build up our account of what typically happens, we will use the experience of an engineering student as he learns the ropes during a placement. Although the placement period in question is only 6 months, it must not be forgotten that the time usually required is much longer, about 2 or 3 years.

We will follow the work of an engineering student during his placement in a CERN (European Organization for Nuclear Research) design office in Geneva. For this student, the difference between what he learned at school and the way things really are in the design office is accentuated by the fact that his assignment seems to be quite simple: define the shape and dimensions of an object, and the materials to be used, so as to meet the specifications of the order givers and the laws of science and nature. Furthermore, this assignment is a good opportunity for the student to apply some CAD (computer-aided design) tools to a real case. The problem does not look complicated at the outset; our student needs only the initial data (the specifications defined by the order givers), a computer console, and a methodical approach.

However, what the young engineer will discover during his placement is that his pre-evaluation of it was much too simple and limited. To be able to fulfill his mission, he will have to change his views and his approach little by little. He will have to rework his initial impression of the design work. He imagined himself sitting in front of his computer designing an object (a scene that is consistently reproduced in literature on design methods). In fact he discovers a social world of varying shapes and sizes. He thought he would have to implement certain methods and apply certain cognitive processes. In fact he finds himself having to negotiate and settle on compromises. He thought the procedure to be followed would be straightforward, starting with the specifications, but everything is complicated by new requirements defined by the order givers following the draft of a first solution. Indeed, the story we are about to tell concerns not only the design of a technical object but also the re-design of an apprentice engineer.

A Strange Supervisory Board

Many young engineers have probably discovered the same thing when starting out on their careers. Few of them, however, have had to deal with the same kind of supervisors as this student. What is more, the specific framework in which the work is done should be underlined. It provides the opportunity to discuss and analyze the trainee engineer's experience and find the terms to express what he sees and feels. The framework therefore has a lot to do with what the young engineer experiences.

His mechanical engineering studies are coming to an end when one of his lecturers, Jean-François Boujut, talks about the possibility of his pursuing a DEA[3] or even a doctorate. Involving research work, the DEA gives students an additional non-technical skill. It is also, the lecturer explains, an opportunity to step back from the operational work required by the

PFE. But the most surprising announcement is that the proposed subject is to be co-supervised by a sociologist. The student is interested to discover that the mechanical engineering teams and the sociology teams in Grenoble are used to working together. However, since our student has only devoted himself to mechanical engineering throughout the course of his studies, he prefers to concentrate on this area and the technical work in hand at the beginning of his placement period. His tutors nevertheless ask him to take an observer's view of the project and closely follow the design process and the actions and interactions that it generates. To begin with, the trainee thinks that his observations are unrelated to his work as a designer. They involve a different part of his mission. This part is non-technical, and the trainee cannot really see what the aim of making such observations is.

The student, placed in a CERN design office in Geneva, is put under the direct responsibility of the head of the office, Bertrand Nicquevert. To the student's great surprise, his engineering school tutors and his "industrial" tutor seem to work hand in hand. They apparently get on really well and share the same opinion of the work he has been given. They say that it is an interesting opportunity to decode and analyze the design process. The student also discovers that Nicquevert holds a master's degree in philosophy. Not your usual mechanical engineer! And as if his supervisors were not an unlikely bunch as it was, Pascal Lécaille— an anthropologist writing a thesis on simulation tools—joins the group in one of the first supervision meetings.

The young engineer can only explain this strange group of supervisors by the interest they have in the other part of his mission, i.e., the social aspects and all the other factors surrounding the actual design work: the language barrier and the cultural differences of the people in the design office, the different age groups and the probable consequences of people retiring, and, finally, the behavioral and relational problems of the office personnel, especially the more senior designers with respect to their young manager. This set-up, in which the social factors are peripheral, external, or simply tacked onto the technical job in hand, is not to be called into question. However, as the design work progresses, we will discover a different way of looking at things, based on the people concerned, the way they react, and their different relationships. Indeed, the problem can only be defined, and a solution found, if these elements are taken into account. Hence, the sociologist's view of the mission does not fall entirely outside the scope of the technical work; it is up to him to try and understand the dynamics of the technical work.

A Simple Object in a Complex Environment

And so, fresh from his mechanical engineering school, the student settles down in the open-plan office. He is given a work surface and a computer console, like the other fifteen office members. With a mission to fulfill and a place to do it in, he thinks that he will be able to get along fine. At school he has learned to use the models, the catalogues of technical solutions, and the appropriate methods for each design phase. With all this learning under his belt, he should have no difficulty finding the right solution to the problem, making the calculations, and checking his work.

What is more, the technical part that he has to design is very simple. It is a wall, or more specifically a shielding disk, to separate two parts inside the ATLAS particle detector. On one side is the calorimeter (for measuring particle energy); on the other is a superconducting magnet. The shielding must prevent all particles other than muons from interfering with the measurement of the trajectographs called "muon chambers." The shielding must absorb photons and neutrons. For this, materials with high absorption rates for such particles (such as polyethylene or copper) are used.

To get to the bottom of the problem, the student starts by reading up about the entire system in which these shielding disks are to be placed. For a week, he concentrates on learning the technical terms relating to the detector. Using a document referred to as the Product Breakdown Structure, he identifies each of the parts, its name, its abbreviation, and its dimensions. He makes several sketches in his logbook for future reference. In doing this, he discovers just how complex the detector really is. He also discovers its impressive size and weight: 25 meters high, 40 meters long, 10,000 tonnes. The detector is to be used to determine the identities, energies, and directions of the particles produced during frontal collisions with proton beams. It is made up of detection and measuring instruments (a trajectograph, a calorimeter, a muon spectrometer), confinement and regulation parts (superconducting coils, cryostats), a range of electronic systems, and various supporting and structural parts.

Designing such a detector obviously involves a large number of people, institutions, and countries. In all, 1,700 physicists and engineers, some of whom can be considered "order givers," are taking part in the project. The "order givers" are the future users. Each element is being designed by a specific team. The CERN design office is one of these teams. As the head of the design office gives a quick overview of the ATLAS detector, he points to different parts of a technical drawing, saying "This is us here, and that's a team in England over there." Working in partnership with others is difficult even if it is routine. The whole thing requires complex coordination among senior managers within CERN and among various project steering committees. The technical complexity of the detector is thus matched by the organizational complexity of its coordination and that of the technical information system.

As the engineering student listens to the explanations offered by the head of the design office, he discovers that coordination among the designers and with the physicists is, in general, a central issue. Far from being a relatively closed space, the design office has a tight working relationship with numerous people from various institutions. The head of the design office talks especially about two categories of partners: the safety department (which is in charge of checking all the calculations for the sensitive parts) and the physicists (who are seen as dreamy idealists always wanting to go one step further without taking into account how feasible their ideas are, or at least that's what it looks like). The design office sometimes relies on the former to temper the wishes of the latter. Even within the office, the question of how to work together is often raised in connection with people's cultural differences and differences in age, and also in relation to where they are seated in the office. Indeed, in the middle of the office there is a row of large cabinets in which 30 years' drawings are kept. This row of cupboards physically divides the office in two. However, in the middle of the row there is a gap about half a meter wide. The head of the design office says: "See that? I've made a space between the cupboards. It was like bringing down the Berlin Wall." This witty remark goes some way toward explaining the reluctance of certain designers to cross the office.

Normally the student would not be concerned with all these coordination problems. The shielding disk that he has to design is just one small part of the whole assembly. He should be able to deliver a detailed draft design of the disk within 3 months. What is more, people hardly seem to be interested in this part. The physicists, for example, have turned their attention to other parts of the detector. The shielding is not seen as a "noble component," says the head of the design office, unlike the particle detectors. It is one of the "common components"—parts that go between instruments to accommodate fluids (e.g. for cooling), cables, and support structures. Indeed, if it were possible, the physicists would like these parts to disappear altogether. For them, cooling should be immediate and homogeneous without having to bother with all the tubing and extras. As for the framework supporting the instruments, this really takes up far too much room. These "common" parts are seen as cumbersome by the physicists who will be using the detector. The designers, however, covet these commons because they impose themselves as constraining boundaries. Of course, when a designer needs just a few extra millimeters, he takes them. The problem is that he is not the only one to do so. In fact, it is the head of the design office who is in charge of making sure all these parts fit together and who has to bring these coveted boundary areas of the system apparatus into existence.

The shielding disk on which the engineering student is working is one secondary part that nobody is really interested in. This is why he is under the impression that he need only analyze the problem and find the solution, without having to get into lengthy negotiations with the physicists. To him, the problem is purely technical. From the outset he knows that the space available for the disk's external dimensions is limited by the external dimensions of the surrounding parts. Using the drawings given to him by his colleagues in the design office, he studies these surrounding parts and their dimensions. He takes into account a few general rules relating to safety and ergonomics, so that the detector can be accessed for maintenance. Thus, his scope of action is limited by a multitude of specifications and requirements imposed by various people involved in the project.

Interactions between Objects

During the first days of his placement period, the head of the design office takes the young engineer around the various departments. He is introduced to many people, some of whom are working on issues directly linked to his own study, some of whom are not. He also takes part in technical coordination meetings that deal with project planning, fitting the various parts together, and the safety of their design. He feels that such meetings are just a matter of procedure.[4] Their aim is above all to check how the project is going and swap information. However, the people in the meetings argue about dates and about documents that haven't been handed over. This has nothing to do with the technical side of the project. It has to do with how projects and meetings are organized and managed.

However, as they go through the corridors from one department to the next, the trainee and his tutor come across various people with whom the tutor strikes up conversation about details regarding various projects that he is in charge of and which he needs to keep in mind. The student is astonished to see that part of the project's technical coordination takes place in the corridors. Like the canteen, the corridors are used as a forum for solving many problems.

The more people they meet, the bigger the trainee's list of contacts becomes, although so many names frighten him. The words of his tutor hardly reassure him. He thought he had come to carry out an engineering assignment first and foremost and, as a kind of sideline, act as an observer. But he discovers he actually has to communicate information, get answers to questions, and negotiate. He discovers then why he has been put in charge of designing the shielding. In fact his tutor had known that it would not be easy. He had even said as much right from the start but the young engineer, judging from the simplicity of the object to be designed, had put this to the back of his mind. When the head of the design office had agreed to take on the student along with his strange group of university supervisors, mechanical engineers rubbing shoulders with sociologists, it was because he thought that an outside view of the situation would help the head of the design office to understand what was in play.

And so the young engineer discovers that his poor shielding disk is the object of important stakes in terms of its functional definition. Indeed, it has to fulfill two functions: to stop particles and to bear the weight of the muon chambers. The problem is that the shielding is surrounded by various neighbors that have to be taken into account.

Moreover, the word 'neighbor' is used both to talk about neighboring physical objects and to refer to the designers of such objects or the order givers. This is why people talk about negotiating with the neighbor when talking about the cryostat, for example. The number of neighbors involved is already quite considerable: several types of muon chambers, the vacuum chamber, the tile calorimeter, the cryostat, the toroid, the rails on which the system has to run, and the electronic data capture boxes. The trainee discovers, for example, that the electronics engineers in charge of designing the data capture systems have designed an enclosure that is too big and have thus reduced the shielding designer's room for maneuvering. In fact, he needs a clearance margin, as it is difficult to know the exact dimensions of the parts once they have been built. If he can't have the data capture box redesigned or moved, the trainee will have to plan a cutout in the shielding disk. And the physicists will probably not like this. Furthermore, it will reduce the disk's rigidity. As the shielding is at the center of a series of neighboring relations, it is an intermediate object; thus, the young engineer has to argue his case if he wants to get the amount of space he requires.

Ten or so neighboring objects mean ten or so teams or individuals to be contacted for data and information concerning their parts, along with all the associated constraints. On the other hand, the trainee discovers that he has to validate his own designs with these people. Some of them work in the design office, but not all of them. Sometimes the trainee has a CERN physicist with a listening ear to deal with; at other times he must deal with a renowned Italian physicist who is impossible to find, or with a Parisian team that does not answer his e-mail requests for information.

Technical or Strategic Work?

The trainee also learns that the work on shielding is strategic for the design office. Indeed, some of the technical parts, such as the muon chamber, are already the objects of dimensioning studies by work groups such as the Muon Layout Collaboration. In order to define the job at hand, the trainee bases his studies on a technical design report drafted by the Muon Collaboration. This document lists the technical features of the chambers and all the teams working on them. It defines the part of the enclosure that concerns him and in which the disk has to fit. Thus, for some of the parts, things have already been defined, and it would be difficult for the trainee to change them. This means that the design office has less room for maneuvering in the design of the structural parts of which it is in charge. The trainee realizes that his tutor has chosen this moment in the project to assign him to the shielding job so that they can have their say in the matter as early as possible. It is essential for the design office not to have to work with a part that is already joined to the rest. The office therefore has to fight to preserve a certain amount of freedom in its design work. Relations between the design office and its partners are as important as all the problems relating to borders, space, and margins.

It is only at this point that the trainee realizes how poorly prepared he was for this situation. He does not really know what kind of attitude he should adopt in this complex social world of hierarchies (related to the organization itself but also to the reputation of people), divisions, and territorial occupations. Therefore, for several weeks the trainee has put these facts on the back burner, preferring to concentrate on what he can do best: a technical job performed at a computer console. He has memorized the environment of the shielding disk from the drawings given to him by a neighbor at the office. He has redefined the enclosures so that he can accurately assess the space available and the margins for maneuvering. And he certainly needs these margins to be able to reinforce the shielding disk so that it can withstand the weight of the muon chambers. Finally, having concentrated only on the technical aspects, he has learned to know where he stands from a technical point of view; thus, he has developed a line of arguments to use with his neighbors if ever he should have to negotiate.

So he beavers away at his computer console, designing, imagining, and calculating. He checks the possibilities for adding reinforcements without disturbing the layout of the muon chambers. These reinforcements are necessary to prevent the disks from buckling under the weight of the chambers. After talking to his office neighbor, who is in charge of integrating the chambers, he drafts some ideas for fixing them to the disk. He discovers that this is the most delicate part of his own design work, as the loads to be borne are considerable and he has little space for adding the framework. Although two-dimensional design software would give him a good idea of the surrounding space available, it does not really give him an overall view.

Having worked with three-dimensional simulation and viewing tools at his engineering school, he decides to use his training period to put one of the software programs to the test. Using it, he is able to show how the muon chambers and shielding disks fit together. (His mechanical engineering tutors are interested but not entirely convinced. They prefer working with concrete analyses rather than such calculation tools.) Next, the trainee designs a framework able to fit into the space available. He simulates various calculations of the framework with different diameters and materials so that he can get a realistic idea of the mechanical stress. He discovers that the framework will not be rigid enough unless it is closed at both ends.

For several weeks, he concentrates on the design of this framework, working in an environment that seems increasingly restrictive and hostile owing to the dimensional constraints and the problems of accessibility: little space available, the need to leave clearance for the detectors to be opened, and the overall suitability of the assembly. There are so many geometric constraints that his first concern is to find a solution that is able to fit inside the space available. While devoting all his time and energy to this problem, he is also able to build up good professional relations with his colleagues in the office. He discovers that everyone there has had to forge a place for himself. The space-related problems of everyone he meets when working on its project are reflected in the design office itself.

Having discovered the importance of "neighbors" when working on a design issue, our student undertakes to list them all, both the technical parts and the people working on them, along with the questions that he would like to ask them. The list is long indeed, and it gets longer since the very notion of neighborhood has to be revised. Before this, it was defined in terms of spatial proximity. It referred to "elements" that may or may not be in (physical) contact with the disk, but are not separated by another element. Perhaps it is his mechanical engineering background that has so far restricted his field of vision. There are in fact several kinds of relations among objects; geographical proximity is not the only one. Indeed, radiation goes through various parts of the detector along with heat, magnetic fields, vibrations (e.g. earthquakes), and gravity. (The detector may be symmetrical, but gravity does not see it that way.) The toroids in the detector generate an intense magnetic field that tends to cause the elements to come together. The magnetic field exerts a force and then checks whether this force affects it or not (or rather whether it affects the shielding). Added to this is the question of maintenance access to the detector. All these forms of interaction can bring distant elements within the system in contact with one another. Drawing up a list of these elements along with the people working on them seems to be the only way forward. In doing this, the trainee is in fact trying to identify and target all the neighbors with the biggest influence on his design work. Defining each one's territorial position (who does what and up to what point) now seems increasingly important to the trainee, as it will allow him to define his own technical work. Moreover, the breakdown of roles played does not now seem as clear as when he started. And so, having worked hard on the technical side of things, the student discovers how essential it is to be able to socially decode relations if he wants to complete his mission successfully. In other words, he has to ask himself who does what and how far is it possible to negotiate. He finds out that certain elements cannot be negotiated, as changing them would put them back on the drawing board. Thus, the trainee comes to analyze the interactions between people, the recurrent nature of certain practices, the rules applied, and the possible interference of all this in his work.

Stabilizing What the Neighbors Want

The trainee also comes to realize that the demands of each person involved are not always clear and are far from stable. The shielding is supposed to fulfill two functions, but when he takes the various neighbors into account he realizes that things are much more complicated. Each neighbor has his own expectations and requirements, only some of which have been put down on paper. What is more, under the instigation of the head of the design office, the trainee has begun a functional analysis.[5] This requires listing and quantifying all the functional features of the disk, with the aim of discontinuing to work on assumptions alone. For example, the physicists say that the shielding should be 100 millimeters thick and made of iron, with a copper base. But why? Which physicist decided that? And on what basis? Where are the data that led these physicists to give such specifications 3 years earlier? Would they say the same thing today? In fact, all these so-called constraints have to be studied again, and the people who defined them must be found and asked why they said what they said. It is no longer possible, at this stage, to continue to rely on the available technical data. It would be better to find out the logical reasoning behind the orders given and whether it is possible to re-negotiate. For example, just how far can the basic functions of the detector be revised? What seemed to be finalized is perhaps not. And so, after 4 months, after the student has done his design work on the supports for the muon chamber, the physicists decide that the way the chambers are mounted does not satisfy them. After viewing the assembly, they realize the need for maintenance access. The support function thus becomes even more complex, requiring the addition of a new structure that is mobile in relation to the disk. All the design work on the direct support of the chambers has to be reviewed. The concept of the mobile structure and that of its supporting copper base have to be validated at the same time. And yet, the young engineer has already spent several weeks and much energy finding a solution. Bringing to light a new element has led to a whole array of fresh constraints calling into question the initial concept. The trainee begins to wonder what he can base his work on. On top of this, he discovers that certain neighbors have taken up more space than was planned simply because they were not aware that neighboring elements had to be taken into account. As far as they were concerned, their elements were surrounded by emptiness. It is easier for members of the design office with the job of integrating the different elements to understand these "neighborhood relations" than for a subcontractor of a distant part.

Now that the neighbors have been identified, the next problem is getting them to talk. Would it be possible, and enough, just to get them around a table? Some of them come to CERN only once a month. Of course, our trainee engineer has neither the power nor the authority to convene them in a meeting. The head of the design office does not have authority to do so either, in view of the number of people involved in the project.[6] So the young engineer decides that the best thing to do is define a certain number of elements himself and draw up the specifications that the physicists should have drawn up in the first place. These would then be submitted with the aim of getting them to react and thus define their needs. It is at this point that he discovers how foreign the "culture of a specifications sheet" (Bertrand Nicquevert's words) is to physicists. They shy away from the idea. They think that if they put their expectations down in writing they will no longer have the power to change them. For the head of the design office, on the other hand, the act of writing them down will force the physicists to express their needs, even if they will have to be modified later. If this is not done, they will be defining a solution to a problem whose terms are unknown.

To begin with, the design work consists in studying each problem one after another. Little by little, the idea that it is necessary to have an overall view, and not just a technical one, emerges. Different people work on each technical element, and it is essential to know exactly what they want and how far it is possible to negotiate with them.

The trainee thus submits his solutions to his neighbors. The drawing of the disk is sent to a physicist so that, through simulation, he can check whether it is acceptable in relation to his needs (i.e., particle absorption). A proposal for modifying the cryostat cover is faxed to the Orsay team in charge of its design. Within the design office, showing drafts of drawings to different colleagues during lunch or in an informal context produces some interesting reactions. It allows the young engineer to see that work with each partner is carried out differently and requires different approaches. At times, the design proposed is provisional, insofar as an unhurried colleague is expected to provide some data. At other times, the engineer has to wait for a reaction to the proposed modifications to a particular part. The assistant technical coordination manager is soon to leave for the United States for a meeting where he should have a chance to raise the question of the muon chamber. A file has to be prepared for him, and he has to be persuaded to bring up the matter. The problem here is that he comes to the office only once every 2 months. Negotiations depend on mediators whose logical approach is not always fully understood by the members of the design office. For some neighboring elements (such as the calorimeter), negotiation is easier, as the colleagues involved work at the Geneva site. For each neighboring element, and hence for each neighbor, there is a specific coordination procedure. In this way, the young engineer comes to understand the interest of the head of the design office in having an outside view of the situation.

Finally, it is interesting to see how simply working on an element such as the shielding draws attention. It has become a subject of interest, enabling questions to be raised sooner rather than later when later would in fact be too late. The person in charge of the muons (a physicist and a close guardian of the muon spectrometer specifications) did not want to get involved in things to do with "services" or "common parts." And yet the design office needs answers to a certain number of questions. Indeed, the more questions it asks the more seriously it will be taken. If no one bothered to do this, the shielding would just become a kind of black hole, a bin for all the neighbors to throw their unresolved problems into. After all, the designers are bound to find a solution later on. Having questions raised with the project actors was, in fact, one of the objectives of the head of the design office when he took on the trainee. He later explained his numerous expectations in relation to his various responsibilities as follows:

As project engineer for the traction system, I didn't have anybody to argue for the shielding disk. Doing the design within the design office was going to enable me to monitor its compatibility with the entire assembly.

As a member of the ATLAS Technical Coordination team, where I am responsible for mechanical integration, the shielding disk presents a number of unlikely neighbors (some of which were only discovered through Grégoire's work). It is at the center of numerous problems but there is nobody to deal with them and the initial design plan was far too succinct. It was essential to have somebody prepared to dig further into this design. But most of all, from a sociological point of view, it was the ideal opportunity to study the dynamics of the design process through a physical experiment.

As a mechanical engineer, and one that works at CERN, there were several small mechanical challenges: calculating the disk, the support, etc. But what I hadn't banked on was that this object would take on a new life, thanks to the initiative taken and the work put in by Grégoire. One of the amusing consequences of this situation is that today I am being offered the responsibility of the shielding disk as project engineer, which is something I wasn't expecting to begin with.

As a philosopher/epistemologist, my questioning centers around technical issues. . . . There are many scientists who like to dapple in philosophy, but there are significantly fewer engineers. The latter are "much more aware of the material aspects of a technical issue" (O. Lavoisy). Following Galison's example, I'm hoping to be able to go further into the question of relations between theoretical physicists and experimental physicists but on the triple basis of "theory/experiment/instrument" as opposed to the traditional epistemological dual basis. Furthermore, using an engineer studying for a DEA, supervised by a human sciences committee, was the ideal opportunity for understanding this area in a much more structured way. It is also an opportunity for me to understand my own work as an engineer and what is being done in the area of How Experiments Begin?[7]

The work of the design engineer turned out to be considerably different from what the young engineer had imagined. He had thought that it
was just a matter of finding the right solution to the problem in hand by
applying the models and methods he had learned in the course of his
studies. He knew from the start that he wouldn't be able to base all his
work on these existing methods and that he would have to invent new
ones, but he certainly didn't think he would have to go so far.

Operational Summary

1. Design work is complex, even for a simple object.   Designing a technical
part, however simple, can quickly prove to be complex when the part in
question lies at the center of a whole system and is linked to a certain
number of other technical parts.

*2. The design work builds up around a network of relations among technical
parts.*   Designing an object involves taking into account a series of other
objects, which are not always in direct contact with this object. These are
related to one another; however, the way they are related is not always
known at the start, and does not necessarily become clear during the
design process. To define the specifications, the designer must describe
this network of relations among technical elements and must go through
it regularly to check on changes made.

3. Objects and their relations are linked to people and social groups. These
objects can be taken into account only if the designer knows these people
or groups (i.e. who orders, who designs, and who uses), their relations,
and the logic behind what they decide and what they do. Of course, this
demands precise attention and a decoding ability to prevent judgment
from being based on simplistic analyses at the beginning of the placement (saying that problems are due to people or technical ideals, etc.).
There are actors behind each technical element, and they act as spokes-
persons for these elements. The elements are at the center of these
people's interest.

4. It is not always clear at the beginning what all the constraints are.   They
are gradually revealed as their relations with other elements are
explored. It is not possible to have them at the start, notably because the
actors themselves do not really know them. The process of designing
solutions and making them viable through drawings leads the actors concerned (or order givers) to talk about the requirements that they would
like to see fulfilled. Bringing the intermediate objects into existence is

therefore an important step that will help the actors to express their needs.

5. It cannot be taken for granted that requirements and technical data are given objectively. In other words, final judgment must be withheld, as the information available may be misleading. It is better to understand how the data are put together (socially and technically) and then regularly test how stable they are.

6. Showing interest in an object gives it life. Working on it, drawing it, and circulating the drawing helps to awaken the interest of the various people involved, to position work in relation to it, and to demonstrate responsibility for it. It also helps those who draw up formal requirements. If no one is interested in an object, it cannot live.

7. To manage relations between technical elements, the designer has to take into account how the actors react and behave in relation to their specific element. Taking into account people and groups means first of all examining how they act, both socially and physically, especially when they have to interact with others.

8. Doing technical work is just one strategy among others. Concentrating on "technical" work, such as entering and processing information using calculation software and CAD, is sometimes seen as a good strategy that can help the designer to report on the situation, his position, what he would reasonably like to obtain from his colleagues, and his margins for maneuvering.

9. Industrial design stimulates discussion. Industrial design is not only a technical means of viewing objects during their design; it also stimulates discussion between designers and other project partners.

2

The Nature and the Stakes of a Tool: The Genesis of a Design Aid Tool for Mechanical Engineering
Olivier Lavoisy

To get a clearer understanding of what is at stake in design offices and in the use of design tools, I observed and participated in the development of one of these computer tools (an OI3C, standing for *outil informatique du calcul de calculs de chaînes de cotes*—in English, computer tool for calculation of chains) in a company manufacturing electrical devices. In this chapter, instead of answering abstract questions such as "What is design?" and "What is CAD?"[1] I shall describe concrete practices connected to design, to objects used, and to interactions between human beings and between other elements. In line with the rules about sciences laid down by Bruno Latour and Steve Woolgar (1988, pp. 23–24), my inquiry aims to grasp the techniques and discourses of specialists so as to become familiar with their production, then to return and express what exactly it is they are doing in a different language. Therefore, this chapter is about working as an ethnographer on techniques—i.e., getting to know a specific field while keeping a certain distance. In doing this, I plan to answer a statement formulated in 1979: "Hundreds of ethnologists have visited every imaginable tribe, penetrated into the deep forests, listed the most exotic customs, photographed and documented domestic relations or the most complex cults. And yet, our industry, our technique, our science, our administration are still not well studied." (Latour and Woolgar 1988, pp. 15–16)

Technical objects contain a part of human reality, and the desire to study human works cannot be separated from an understanding of the technique used. As Georges Simondon noted (1989, p. 9): "The apparent opposition between culture and technique, between man and machine, is false and unfounded; behind it lies only ignorance and resentment. This all too easy humanism masks a reality rich in human effort and natural forces, essential elements making up the world of technical objects, themselves mediators between nature and man."

There are various objects at stake in the design of the OI3C tool. There is in fact a blend of many things, notably the identity of the firm, the firm's training policy, and the philosophy on which the design process is based. For example, given the company's new identity (I'll call it Green, since it is the result of a merger of two companies I'll call Blue and Yellow), we can raise these questions: Has the merger really taken place? Is the creation of a new industrial entity just a matter of changing the organizational chart and the logo, or finding a joint name, or moving employees around? Isn't identity also a question of objects and methods of work in computer programs? To support this theory, we shall look at OI3C from several vantage points.

We shall first see how the OI3C computer tool becomes the core of a diverse network. The discovery of the network will go hand in hand with the opening of a black box, the very heart (called the *solver*) of this tool. Next we shall consider this central activity, referred to by the actors as "solver validation." We shall then examine how OI3C, as a design tool aid, can also be used as an instrument of coordination. OI3C will finally be situated in the dynamics of standardization of the company thanks to the part of the tool called "the dimensions manager." This analysis will lead to a discussion of graphic representation.

Let us start by studying the "technical" population of the Green firm: the Standardization and Technical Coordination (hereafter STC) department[2] and a certain number of design offices in the Green firm. Four people, referred to hereafter as *project developers* or *OI3C genitors*, are at the heart of the action: the project coordinator, the trainee (me), the trainee's industrial supervisor (the second project coordinator), and the computer specialist (an employee of an outside firm).

OI3C: The Core of the Network

One way of understanding the OI3C software is to describe it and the objects to which it is linked. The aim is to bring it out from backstage and put it under the spotlight to see it in action and to discover it through what people say and do as they use it.

It is difficult to enter a new field and become interested in it if certain elements make no sense at all. Any observer sees what he is prepared to see, or what particularly surprises him in relation to what he already knows. Observation is structured by our view of things. Inside the STC department, the birthplace of OI3C, an observer who is uninformed about the project and its context would merely see people sitting at their desks from dawn till dusk, staring at colors moving on a computer screen. A bell often rings and someone starts speaking on the phone. What else could the observer say about such a scene? How is he to make sense of it? How can he guess that these people sitting there are paid to do so and that they are all highly qualified? Although simple, such a description assumes that the observer already knows what computers and telephones are. Indeed, we all share some knowledge about and some references to their use. For example, both the observer and ourselves know that there is another person at the other end of the line and that the individual holding the receiver is not just speaking to an inanimate object in a fit of madness.

Software, Paper, and Telecommunications

Geographically, OI3C is situated in a room large enough to hold about ten people. A computer workstation sits on a table. On the screen, in a colored rectangle, lines, arrows, and figures can be seen. The rectangle is labeled "Pro/ENGINEER"; this is the name of the CAD software. The screen is labeled "Sun," indicating its brand. The keyboard is a QWERTY model of the Anglo-Saxon type, not the AZERTY type even though we are in France. (The significance of this will become evident below.) Where is OI3C? It doesn't exactly jump out at us. Only a trained eye will notice the name "OI3C" among the others on the screen. So far OI3C is just a name (of a software) appearing on the screen when the PRO/ENGINEER CAD software is running on a workstation situated in an office next to other similar offices within the STC department in which 20 people work.

The network surrounding OI3C begins to emerge when you follow the conversations between people. It happens gradually as you watch a member of the STC department, sitting in front of his workstation with his telephone to hand, communicating regularly with software developers outside Green, or with the members of various design departments within Green. A fax machine is being used to send and receive texts and sketches, and during phone conversations the coordinator of the OI3C project often looks at faxed documents. Computer files circulate between computers and printers.

Thus, OI3C is not only a reference on a computer screen. One can see signs of it on printed sheets, on computer program pages, in configuration files, and in sketches. It is situated in a specific area of the department where there are two photocopiers (one of which is shared with the other department in the building), a printer connected to the network, a portable printer (which is moved from office to office), and two printers connected to an independent computer.

Such observations give the impression that this equipment is strategically placed so as to organize the movement of department employees. The room in which the blank paper and the printer connected to the network are located is the same distance from one end of the corridor as from the other. Along this corridor are the doors to the department offices, and at either end of the corridor a photocopier is stationed. Each person also has a telephone, so there are two or three per office. Much like the coffee machine, these spaces seem to encourage discussions (e.g., in regard to a sheet that has just come out of the printer: "Hey, look at this!"). These are, in fact, the only places in which one is not "in someone else's territory."

Using a fax machine requires more physical movement than using a telephone. First one must produce the medium (a printed or handwritten sheet, or a photocopy); then—and this is especially true in this case—one must move around the building (the STC department is on the first floor; the fax machine is on the ground floor, next to the receptionist). Indeed, the departments that share the fax machine occupy three entire floors. Nevertheless, the project coordinator does not always go downstairs. He can also send a document from his personal computer (but not from the workstation) by means of the internal e-mail service. And there is also a system for internal mail, which is delivered to every office twice a day. The topography of the department is an important element. OI3C can be worked on from one's own office (by phone, internal electronic mail, or internal mail), or on the workstation that hosts it (where there is yet another telephone), or by fax. However, choosing the right piece of equipment does not depend only on whether one feels like moving. Observation teaches us that the most determining factor is the behavior of the correspondent outside the department. If (as is the norm among engineers) he does not often consult his electronic mail, it is better to use the fax machine (and, in most cases, the telephone too) to warn of the arrival or departure of a document, or to discuss a document in real time.

Paper is very present in this universe. On the basis of an estimate made by the person in charge of paper supplies, the design office consumes about nine boxes of paper—that is, 22,500 sheets—every 2 months. Since there are 20 people and the number of working days in a month is about 20, each person uses about 30 sheets a day. Indeed, the department from which OI3C emerged can almost be described as a factory producing printed paper, especially when one considers that the offices are filled with files and documentation. This department produces texts and, to a lesser extent, sketches and computer files. It is therefore neither a design office nor a calculation center. And, as its abbreviation suggests, it is in charge of standardization and technical coordination for Green. It is through the various media created that OI3C actually exists, since the software is not restricted to the host computer alone (even if at first glance it seems to be).

OI3C's History

The start of my inquiry (that is, my reading of documents and my interviews) revealed two fundamental aspects from the very start. First, a need for confidentiality was often expressed—particularly at the beginning of my stay, when mutual trust was only just emerging. This underlines that the new tool is considered from the beginning as a strategic element lending a competitive edge to the firm. Second, the organization set up to create OI3C is project based, with a project coordinator, correspondents from other design offices that will be using OI3C in the future, and an outside computer specialist.

The story of the genesis of OI3C told by the project manager is very straightforward. First, a group of people come together, and "user need specifications" are defined by a sample group of future users. "Functional specifications" are then defined by this same group for the benefit of the computer scientist; these are then developed by the group until there is a model of the OI3C tool; next, tests are performed by the project coordinator, the trainee, and the users (the latter being referred to as "pilots"). Finally, an industrial ("1.0") version, to be installed at sites belonging to various divisions of Green that participated in the financing of the project, is released.

However, dividing the proceedings up into such clear-cut time sequences is too easy. It is necessary to analyze the periods in the project's evolution objectively (Prost 1996, p. 114). Discovering the history of OI3C actually consists in specifying periods that make sense—in other words, "substituting a significant structure for the imperceptible continuance of time" (ibid., p. 115).

First period: suspicion

Before 1994, the designers of the late Yellow firm became aware of the limits of their tools when they realized that, in the years since the introduction of the MEDUSA software, their calculations of chains of dimensions were indissociably bound to this software. Then, Pro/ENGINEER began to take the place of this software and the designers started to formulate "a request . . . for the CAD-CAM[3] department to develop something [autonomous]." At the same time, the designers of the late Blue company began to realize that their tool (called EPURES), which "was not interfaced with the CAD," was "somewhat limited."

New tools were needed but at that time relations between the ex-Blue and ex-Yellow divisions and the central CAD-CAM department were full of suspicion. The ex-Blue divisions felt they were ahead in the preparation of methods, including design, and "were afraid of imposing something on Green which would not meet their need." Besides, before 1994, researchers in the CAD-CAM department carried out their first investigations on commercial tools (e.g., TI TOL, from Texas Instruments) and on an internally developed tool that seemed to offer potential. Their task was to test those tools to see how well they matched the specifications defined. It was in fact the STC department that did these tests when it should have been the CAD-CAM department. This role swap meant that the choice of tool moved from a department responsible for computer resources to a department responsible for defining work methods. One result of this transfer is that when I became involved in the project, in March 1996, nobody in the CAD-CAM department had yet seen any running version of OI3C. They were not involved in the project because, as the members of STC said, "they put tools before the methods and needs of the users" thus undermining interest in the OI3C project by wanting "to put the cart before the horse."

Second period: institutionalization of the project
One of the first documents I unearthed, which dates back to March 17, 1995, emanated from the OI3C project coordinator. This two-page fax plans study group meetings and restates the objective ("to draft the functional specifications of a computer tool for the calculation of chains of dimensions / functional dimensioning and statistical tolerancing"), the deadline (by the end of June 1995), and the composition of the study group (seven people, including the coordinator).

The specifications sheet of the OI3C tool, dated September 19, 1995, defines OI3C by its desired capabilities and by diagrams in which rectangles and circles are used to represent OI3C "beings" (CAD[4] tool, database, and exchanges between these indicated by lines and arrows). These specifications refer to the "specifications of user needs" mentioned in the oldest available internal note (dated February 28, 1995 and drawn up by the project coordinator).

To understand what happened before February 28, 1995, I had to resort to interviews. As is often the case in history, I was faced with a scarcity of documents, and some of those to which I had access were confidential. Why? The OI3C project is seen as a competitive advantage. The warning "not to be distributed outside the Green group" in red felt-tip on the front page of the document "Functional Specifications" is a means of protecting OI3C along with the firm's core skills. The OI3C software thus appears to part and parcel of the firm's collective identity and skills.

"What got everything moving was the decision to make it a Green project," one person from the late Blue division said: by deciding at the beginning of 1995 to create a study group aimed at internal development of a tool to meet the requirements of internal users, the CAD-CAM manager seemed to have removed all opposition. This study group then drafted the OI3C specifications. It was the initial job of the person who was project coordinator in 1996 to represent all the divisions, not just the two main divisions of the late Blue and Yellow firms, and to act as advisor between the two. In actual fact, there were to be two coordinators, the first relying on his design office experience in the extra-high-voltage department at Yellow and the second on his CAD skills (he had been an in-house trainer before coming to the STC department).

What we are seeing here is the phenomenon of project institutionalization during which efforts and actions are crystallized into an organization backed up by rules, procedures, meetings and capitalization of the project. The motive for the project is that the market does not provide a satisfactory tool for the expectations of the designers in terms of autonomy (like EPURES of Blue) and of links with the CAD system (which is what Yellow's designers were used to with MEDUSA). The setting up of the internal study group means that various actors from the institution call upon an outside firm to translate the project into computer form. The notion of "outsideness" is, however, relative: the developer is a former employee of the CAD-CAM department who has founded his own firm. He shares the same memories of the late Yellow with the two coordinators of STC. The manager of the CAD-CAM department did not bother to explore other market possibilities; working with a former colleague who was a Pro/ENGINEER specialist was an obvious choice.

Third period: crystallization of the project into an object
The year 1996 was a very busy one for the project. The first mention of the OI3C programming appears in a summons to a meeting on February 13 where the subject was "promotion of computer development." One month later, I arrived at STC to take part in the solver validation as a trainee.

The volume of documents increases considerably during this period, and new writers begin to emerge. First, the trainees' supervisor organizes a third day of discussions on functional dimensioning; this day is to bring together more than a hundred people from the quality and production engineering departments and from the design offices of various divisions, both French and foreign, for a day of presentations and debates. The subject is "Functional Dimensioning: From Design to Manufacture." Three meetings are scheduled with, each time an invitation together with minutes being sent out, making the list of documents, which I shall refer to as "inscriptions" (texts, diagrams, tables, etc.), even longer. These documents trace the genesis of the functional dimensioning project. Furthermore, the specific job of the trainee is to produce documents for future users and to monitor the test phase. During this period, about twenty documents are drafted, with the aim of determining, motivating, and coordinating the various actors. These document are necessary for the creation of a tangible tool: the OI3C program.

The architecture of the project gives it even more weight. Built up by separate teams who do not always trust one another, it is gradually institutionalized until it crystallizes into a computer model, texts, and strings of data. As I have already mentioned, the difficulty of building a collective identity and a collective activity can be seen through this technical development: being involved in the construction of OI3C is like being part of the construction of the Green firm. For these reasons, the genesis of OI3C can be understood only by plunging this software-to-be into the environment of other objects with which it interacts. Similarly, it develops over time, and all the struggles, tensions, and future speculation focus on it as an object. We also see that the project group and the industrial organization are set up at the same time as the tool.

The next stage is the crucial validation of the solver. It is, of course, at this point that the solver's technical performance is checked. But also checked are whether the object (the software installed, OI3C) is in line with the subject's expectations and needs and whether the industrial merger (i.e., having the same tool for all the designers) is in fact feasible.

An Instrument of Technical Coordination

In the previous section we looked at OI3C from the point of view of a collective of persons and objects. In this section we shall be moving between the STC department and a design office in order to focus on what the last two letters of the French abbreviation STC stand for: technical coordination.

Degree of Contextualization or De-Contextualization of the Test

Officially, OI3C is a computer program whose job is to help designers define the dimensions of the objects they conceive (geometrical dimensioning) using calculations; it is a question of determining a size considered as valid, taking into account various constraints. Thus, OI3C is defined independently of any context of specific usage and is supposed to be universal.

At this stage of the project we enter the test phase. For 6 months, the software will undergo various tests to verify that it is able to meet the objectives, namely to help the designers size parts that are being designed. The goal is to apply supposed universally valid software to different contexts of usage, which, on the contrary, are highly local. Testing it this way makes it possible to check the whole contents of the software and see how effectively the users are able to handle it.

For the validation tests, a number of points taken from the "real practices" of users are identified and tested on the software. The testers use examples and problems actually met by the designers. They compare the solutions (dimensions) produced by OI3C against the numeric results obtained by other software packages considered in the firm as accurate and against the numeric results obtained before and after modification of the OI3C program. Examples taken for comparison are, at first, supplied by "pilot users"; they are supposed to represent situations that the designers would like to be able to work on using OI3C.

Let us follow our testers, who are the two project coordinators, the computer scientist, and the trainee. In practice, they handle elements that no longer have very much technical meaning. The examples presented by the users are de-contextualized as soon as they enter the STC department. Testers, for example, are quite incapable of talking about the part that needs to be defined in terms of size, and about the assembly to which it is to be added. One of these examples is even paradigmatic; it is almost always used for tests, in negotiations, and when people have to be convinced and explanations given. It is disconnected from the object to which it refers but also from the practices and problems of the design office that proposed it as a test. Exactly what they know about the real part can almost be seen in what is presented below, starting with figure 1.

Once de-contextualized, the example is used to confront the software with a given situation based on reality. In other words, the validation is (re-) contextualized. Contextualization continues, moreover, as site visits are made in various design offices.

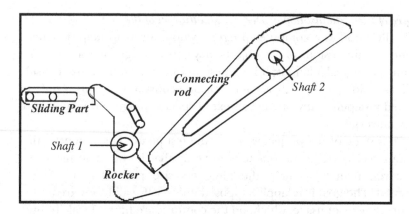

Figure 1

The example is not only that of a part; it is also an example of need expressed by the designers. For testers, this need, also de-contextualized, represents the ability to automatically calculate certain dimensions, the "Operational Suitability Conditions [OSC][5] or Functional Requirements [FR]." It is about dimensions (sizes, distances, and angles) that are important for a product to operate correctly. Operation, of course, does not depend only on a part's dimensions; it depends on many different elements. In this case, however, geometry (dimensions) alone is taken into account; hence the name "computer tool for calculating chains of dimensions."

In this example, one of the distances is assumed to be essential for the part to operate correctly. It is marked by the abbreviation CAE (OSC in English). But why this distance should matter rather than another is a question to which the testers have no answer. The part is then diagram-matically represented using the CAD software PRO/ENGINEER. (See figure 2.) The users of OI3C then have to transform this Pro/ENGINEER diagram into geometrical data, which are entered into the OI3C soft-ware. The part is then represented in the form of a working drawing (fig-ure 3) built up of lines and textual symbols. The "chain of dimensions" (that is, the list of dimensions that influence one another and their respective positions) is then entered in the Pro/ENGINEER software.

OI3C then calculates the "Functional Requirement" dimension by defining two parameters: the "nominal value" (ideal value) and its "inter-val of tolerance" (acceptable precision differences on the real parts; i.e., "Between this value and that, it is valid"). In the language of the OI3C project group, the CAE is the size defined in the Functional Specifica-

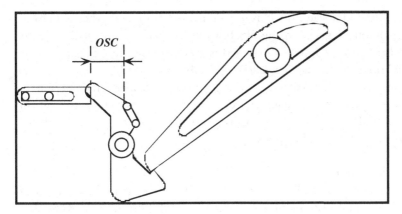

Figure 2

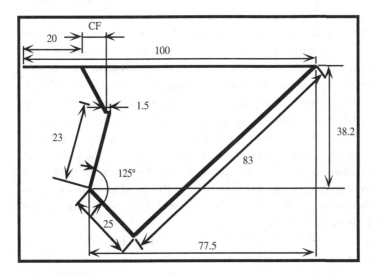

Figure 3

tions, while the functional requirement is the result of the calculation which aims to respect this prescribed CAE value. Ideally, the one should be equal to the other; only the way they are found should differ (constraint expressed in functional specifications by the customer representative vs. the result calculated by the software based on a proposed solution for the customer). To do this, OI3C creates a mathematical object (an equation that is said to be functional). This is then used to calculate the value of the functional requirement using a calculation algorithm chosen by the computer scientist. This in fact involves translating the value from the geometry register to the algebra register. The chain of dimensions then becomes the information used to calculate the "functional requirement" (in French, *condition fonctionnelle*, abbreviated CF).

These new objects (sketch, working drawing, OSC, FR, chain of dimensions, and functional equation) are manipulated by the testers to confirm that the OI3C software is appropriate for the real cases and the needs expressed by those who speak for future users.

Re-Contextualization of the Test

After this first series of tests within the STC department, a second series is carried out on site in a design office that will host the test (β) version of OI3C. As for the description of the STC department, we shall begin by depicting the "design world" that is to host OI3C and the breakdown of objects with which OI3C will have to work.

In this design office (in French, *bureaux d'études*, abbreviated BE), we find photocopiers, reams of A4 paper, faxes, and personal letters, just like in STC. On the desk where the workstation is, there is sometimes a telephone (this site has numerous wireless phones), there are files of standards (notably the "technical standard" of Green), and there are objects that apparently come from the production workshop. In this BE, jobs relating to the graphic work (on computer, rough drawings, etc.) and jobs stemming from data handling seem to be separate. Each designer also has a working desk, often containing a pile of department memos, a diary, internal publications, and a phone (or the fixed base of a wireless phone). This office overflows onto the nearby walls, where there are calendars, family photos, Post-Its, etc.

This visible opposition is less cut and dried than it appears. Indeed, most of the time the designers hover around the workstation, where there is a big plan and a note pad used to make quick sketches. With the wireless phone in one hand and the keyboard in the other, the designer can be seen at his workstation talking with the STC department about the way to use OI3C. Coordination between the BE and STC is structured around OI3C. Between the BE and the workshop, coordination also seems to be via physical models and examples of good or bad manufactured objects—the BE's work tables and bookcases are filled with them. Because the members of the BE move these objects around when they are talking about calculating chains of dimensions, we shall also follow these objects by going back to the workshop. Highly automated, the workshop is devoted to the manufacture of parts and the assembly of small devices (e.g., circuit breakers). How does a "functional distance" become a concrete reality here? During the visit, the BE pilot user takes two plastic parts at random and tries to assemble them. Perhaps he forces them a little, or even too much. After a discussion with the operator, the conclusion is that, yes, the designer has forced the parts. A series of questions then ensues: Has the workshop operator not respected the tolerance indicated on the plan? Were the tolerances put on the plan by the designer inappropriate? Has there been a calculation mistake?

And so we discover two new places and can now consider the physical comings and goings of the people working in them. The BE is part of a rectangular building the length of which is traversed by a corridor separating it into two halves. Between the BE and the workshop the production engineering office is located. One might think that the work mode used here is sequential, but the aisle leading from the BE to the workshop is used in both directions as people and objects regularly go back and forth between the two. So, in the BE, the small manufactured plastic objects taken from the workshop have their equivalents in the form of lines, arcs of a circle, plans, and three-dimensional representations on the workstation's flat screen: plans from the BE head to the workshop, while parts produced in the workshop head in the direction of the BE. Through much mediation, the dimensions worked on in the STC department are bound to the physical parts produced in the workshop, and tracing the route they take from one place to the other helps the OI3C designers to re-contextualize their product. In this way, they come across new needs and situations of use that differ from those encountered in STC. These situations are themselves converted into chains of specific dimensions, providing the designers of OI3C with more food for thought.

The reactions of the pilot user, who is recognized within the department as a "specialist in chains of dimensions," show us how rigid the new tool is and how it creates a certain number of irreversible features. The pilot user works on the EPURES software that OI3C is supposed to replace. Using EPURES means working with circles, whereas OI3C is based on lines (straight segments). How valid OI3C is for this pilot user at this time will depend on the possibilities of adapting it to various modes of representation. Whatever the case, we know nothing about the history of the decision to change from circles to lines. It may have come from the first users consulted, the project manager, or the computer scientist. We can only look at how OI3C is used; theory alone cannot answer such a question. In this respect, our inquiry can go no further. We can only identify one obstacle to validation and discover that by creating compulsory computer passages and methods to be followed—by imposing privileged command sequences—OI3C tends to recommend a line of thought (based on lines rather than circles and on geometrical conditions rather than on mechanical stress, etc.). Finally, what is at stake during the OI3C validation process is not just the software. The process also involves collective production of knowledge and mutual adaptation. OI3C thus represents an opportunity to improve coordination between STC and the design offices—coordination that is at the heart of the validation work. Indeed, to accompany field testing, questionnaires are sent out to the pilot users so that the OI3C genitors can situate the causes of operating errors and become familiar with improvement needs. As it turns out, the questionnaires are never sent back. On the other hand, by telephone and during study group meetings, remarks are made informally and new negotiations started, and thus OI3C moves closer to STC and the various design offices in an unexpected manner.

The Solver Is Not So Black a Box as It May Appear

Let us return now to STC to open the other black box: the solver. There are in fact numerous bends in the road before the OI3C genitors go into the design offices to install the β version to "confirm the solver." There is actually no straight line between "the solver must be validated" and "the solver is validated." The process is dynamic and even unstable. Many things happen along the way. Let us now follow the construction of the solver from the beginning of June to the end of August.

At the beginning of June, everything goes wrong

June 6, 1996, the day on which the pilot users are trained, clearly stands out as the day on which the "solver is validated," just before the β version is installed at their respective sites. Nevertheless, my field notes for June 7 read: "Decision not to install β before fixing Modif.Cote problem [Command allowing a dimension in OI3C to be modified]."

On June 4, the Sun Spark workstation on which the trainee and his supervisor work is replaced by another machine because the "leasing" contract expires. The new "Ultraspark" machine is half the price and is supposed to be twice as fast, even though it is dependent on a server (whereas the previous station was autonomous). The keyboard is a little softer, and the mouse, although it still has three buttons, is more like a PC mouse, with a ball running over a pad instead of an optical device reflecting off a metal plate. A few hours are not enough to change the machine and transfer the files. Furthermore, at the same time, the CAD-CAM department moves from the first floor of the research center (where the STC department is located) to new offices about a kilometer away. The problem is that one of the prerogatives of this department is to install the new versions of the software packages, operating systems, and hardware for CAD. This move takes people and tools away, one of these tools being the tracer that was used by STC to print a working drawing or to create a transparency for a meeting. The presentation of OI3C to the future users has to take place on June 6, with a software version that must work. The computer scientist working on OI3C is summoned to STC to look for the causes of the new operating errors. These may be due to the hardware, the solver, or the configuration files. It's a race against time to present software that works—not prototype software—to the users of every pilot site on June 6.

What do the actors cling to in this unstable universe? "To the set of working drawings," i.e., one pack of sketches previously entered into the OI3C software and used to test it. The working drawings should therefore be used as a reference, but unfortunately the tracer has been moved and the disc player of the new station does not work. How can first-rate working drawings be printed? How can paper versions of the files of chains of dimensions be produced for the invited users? The actors use all their wits transferring the computer files from the CAD system to the text processing system. They make a text out of a graphics file. At the same time, the computer development engineer keeps track of any problems, sitting glued to the computer screen in the midst of a multitude of windows, one of which displays a programming error search program called the "debugger." The various actors instigate many and varied transfers between machines. They measure to what extent the software packages (for example, Pro/ENGINEER) depend on operating contexts (PC vs. workstation). Thus, the main problem encountered by the computer scientist is linked to the fact that he usually works "under an MS-DOS environment, on a PC," while the users of Green work on "UNIX environment" workstations. The solver, which is supposed to be independent and specifically developed for this, must constantly be adapted (to "redefine the configuration files," add finishing touches to certain features, etc.). The project very quickly has not just one solver but two: the solver itself looks decidedly less and less like a hard core.

Finally, on June 6, an OI3C version works, and several working drawings can be created and worked on. The pilot users are delighted, as are the OI3C genitors. However, the day's discussions bring new problems to light. The pilot users prove to be highly sensitive to the possibility of modifying the chains of dimensions without having to start all over for each modification. My field note pad reads "Impressed the audience when Modif.Cote OK" and "Otherwise, don't count on me to sell it." Yet this command has not really been finalized, and on June 7 I note the "decision not to install β before fixing Modif.Cote problem."

Besides, there are still some differences between software packages (EPURES and OI3C) for specific features relating to the calculation results. If there are still some differences, can we really say that the solver has been validated? The answer here is No. On the other hand, we mustn't forget that the validation criteria have changed. The feasibility of the software now depends on whether it can be connected to the CAD software packages. Since the solver works "more or less," the connections are what really matter now. In other terms, the solver seems to be validated!

At the beginning of July, a new crisis
During the summer, the first installations of the β version take place, the first at a site close to STC at the beginning of July, the second toward the middle of the month and about 100 kilometers away, the third at the end of August, and the fourth at the beginning of September (at a site where I was to stay for a week).

During the installation at the first site, on July 4, it is not possible to complete a whole chain of dimensions, even the one that has never posed a problem in the past. But since June 6 the only modification has consisted in verifying the accuracy of results with the chains already drawn: no new chain has been introduced. This new incident obliges the trainee to start again using the printed file containing all the information necessary to define a chain. At the same time, his supervisor is faced with the same problem on the very machines used for the presentation of June 6. During this phase of the test, the main fear of the OI3C genitors is "the unpredictable bug," i.e., the one they do not know how to track as it strikes when least expected. It attacks the simplest chains of dimensions, while the more complex chains do not pose a problem. Why is this? Is it a problem relating to the solver or to the change of machine? The computer scientist eventually discovers that it concerns the solver's calculation precision, which means that the solver is not just a simple tool for calculating functional requirements. It reflects both the dexterity of the developer and the philosophy and decisions of the STC members.

At the end of August, Modif.Cote's meaning is modified

At the end of August, three sites have their own β version of OI3C that works well enough for the pilot users to test the solver "validated" by STC using their own chains of dimensions.

During a visit, some pilot users explain to the OI3C genitors that the command Modif.Cote is, in fact, a communication tool for the design office and the production engineering department. It is important to be able to add a dimension to the chain (for example, stemming from a part added for manufacturing purposes) without starting all the calculations over from scratch. The flexibility of the software seems to be essential for the proper coordination between departments. Nevertheless, after thinking about it for a few days, one of the leaders of the OI3C project comes up with quite a different idea. For him the Modif.Cote command is not very important, since defining a chain of dimensions requires only a few minutes' work. Besides, the command is not easy to finalize, owing to the scarcity of information that Pro/ENGINEER (often presented as a black box) supplies to applications (such as OI3C) that are added to it.

At the end of the validation tests, the solver loses all appearance of being a hard core. Reworked by the development engineer, reviewed by the project manager, then declared valid when in fact it is not completely under control, the solver is not just a tool, independent of all actors and simply waiting to be validated; on the contrary, it evolves along with the interactions into which it enters. Finally, the validated solver is the result of a long validation process rather than the precursor of such a process. Not as difficult as it seemed, it has finally been stabilized after numerous adaptations. It is thus capable of holding together a network of workstations, sets of working drawings, and chains of dimensions, but it also acts as an intermediary among various Blue and Yellow departments and among STC, the design offices, and the workshops. Furthermore, the solver has become a constructive argument for Green through the speeches and written notes of the actors involved. Now properly set up, it joins other practices and is the subject of a new consensus, which contributes to the constitution of Green.

An Instrument of Standardization

The first letter of the abbreviation STC means Standardization. In this section we shall be studying OI3C at the last site where it was installed. The importance of standardization at Green came to light in an almost anecdotal manner. As already mentioned, QWERTY keyboards are used on workstations at STC. The French AZERTY keyboard would have been less surprising, but, after asking the members of the STC department, we discover that this choice is strategic. Indeed, the CAD-CAM department of Green considers that, since the firm is now multinational, tools must be standardized on a common basis. QWERTY keyboards are used "so that anybody can work anywhere in the group." Besides the issues at stake with OI3C, there is a general move toward harmonization and standardization in the firm that is being crystallized into different technical choices. In spite of this, the actors do not unanimously agree on the choices made, as we saw on another site where the designers have AZERTY keyboards. When this was pointed out, our STC colleague was also surprised. The local designer's explanation was quite simply "we are resisting." Indeed, this site deliberately chose AZERTY keyboards to assert its specifically local roots in opposition to the multinational nature of the group. This is not so much opposition between French and English as it is an expression of Blue's resistance to Green. Such standardization/differentiation stakes are at the heart of the technical and semantic choices of the OI3C project. The OI3C tool is probably the first tool—except Pro/ENGINEER, which is a commercial product—intended for use by all the designers of Green. In other words, it is the first tool that is supposed to create a common reference and promote the new design philosophy, centered on customer needs. At the same time, it offers an opportunity to remind the CAD-CAM department, like STC, that tools must be adapted to local contexts. For the moment, OI3C "speaks" in French.

Standardization Means Training

Training creates tension that is all the more intense for two reasons: it is through training that the "company spirit" (*esprit maison*) is promoted, and in this particular case it involves two spirits, that of Blue and that of Yellow. In the 1970s, Green gradually started taking over Yellow; then, in the 1980s, Yellow launched a hostile bid to take over Blue. In 1994 the two companies merged to become Green. The company was entirely recomposed and redefined in terms of activities (centered on electrical technology). After the merger, two internal training centers, one in Paris (formerly Blue's) and one in the south of France (formerly Yellow's), were restructured to promote the same spirit and develop the same skills. Standardization also requires tools and methods. For STC, you cannot have one without the other; tools and methods go hand in hand. Thus, the development of the OI3C tool in the design office (to which STC belongs) is justified as an alternative to its development in the CAD-CAM department.

In this context, training is not just about helping people to overcome their "resistance to change." It is a means of standardizing practices and channeling feedback. It allows the trainers to assess the differences between what is prescribed by STC, for instance, and what they actually see and hear during training courses. Trainers are looked up to as experts. During the courses they run, they are confronted with the experiences of "trainee" designers; at the same time, they are able to promote their own messages. This opportunity to discuss points of view is not, however, inevitably transparent in terms of design practices, because the course attendees only testify to a small part of actual practices. It is only through closer observation that the details can be analyzed, which is why trainees get to stay at an experimental site for a while.

CAD in Use

We see three people around a workstation. One has his right hand on the mouse. The other two, dressed in overalls, have brought a metal prototype. On the screen, we see colored volumes, overlapping lines, and dimensions. On the table are a very large plan of a metal part and a slide gauge. The three people pick up these objects in turn. Using the Pro/ENGINEER software, they seem to be superimposing geometry and text (dimensions) and abstract and realistic views (colored and shaded areas). They manipulate the symbols, the abstractions, and the agreements built into the graphic representations. For them, realism is obviously neither a means of charming the spectator nor a selling point. On the contrary, they use the realistic views as a means of negotiation. As they talk, they manipulate the physical models, the slide gauge, the colored pencils, and the technical drawings. Furthermore, the agreements to which they refer and their competence concerning codes are linked to the standardization of objects and tools. There are two different work spaces for the people in the office to use. The first is the 21-inch screen of the workstation, which has several levels because the various representations of the object are registered in multiple "windows." The second is the A0 paper, which is too large to give an immediate view of the whole information at one glance. These physically different spaces get muddled up in practice instead of acting as substitutes for one another, which is what one might imagine they are there for.

In this forest of objects and representations of objects, OI3C is just one element among others. It introduces a new point of view into a universe

already rich in objects and tools, promoting different design philoso-
phies and involving different constraints. Going from one to another, the
designers simulate assemblies and solutions. "And how about if I tried
that?" can be heard from the man at the workstation from time to time.
Going from one action-based logic to another, he tries to produce a representation that reflects a consensus of opinion from the actors before
him. Seen this way, the standardization project channeled via OI3C can
only be partial. After all, it is just one among others. Its binding force
depends on the multiple elements making up the concrete situation of
each designer. The universality of the software can be seen as relative if
one takes into account the contingency of the various situations through
which it passes, unless these situations aim to achieve a certain equiva-
lence by other means or under the effects of other constraints.

OI3C, an Assistant to Technical Management

In this design office, expressions such as "lead time," "we've got too
much to do," "I'm behind," and "I haven't got time for . . . " are recurrent. These expressions make even more of an impression when we consider that the designers not only spend time designing—a highly valued
task—but also manage plans, documents, and a huge amount of data.
The design office has a lot of cupboards with horizontal drawers moving on rails, numerous metal boxes, tubes containing plans, and tables
overflowing with papers full of text and tables. It is not unheard of for
a sheet of A4 paper, called the "list of dimensions," to be put on an A0
plan next to a workstation. Such lists must be constantly updated to
ensure coherence between the set of documents (texts and plans) corresponding to a product throughout its industrial and commercial life.
This technical management calls into play many actors (the draftsman
is not always the one that makes the modification to a plan, for example), and there is a heavy amount of data to be treated, copied out,
passed on, and remembered. (For example, an electrical equipment
project at Green can contain up to 700 dimensions included in some
300 chains of dimensions.)

To make technical data management successful, the designers use the
photocopier a lot. They copy pre-existing templates, which they then fill
in, complete, and annotate with pens (which they typically carry in their
shirt pockets). During visits to various Green sites, I observed that the
designers use many sheets of A4, which one might have thought would
be used more in administrative departments than in a design office. They
use them to create double-entry tables, for example, with the names of

dimensions and the names of functional requirements. Certain boxes are filled with × marks to indicate that the corresponding dimension is a part of the chain used to fulfill a given functional condition. It is at this point that OI3C should take over to change the way tasks are dealt out, since it incorporates a "dimensions manager," i.e., a complete list of the dimensions of a project. Using this, the tool is supposed to integrate an economic dimension into the design process: there is always a dimension margin, or "tolerance," and the higher the precision required the greater the effect on the manufacturing costs. The OI3C dimensions manager is thus supposed to be a simulation tool, since it calculates the impact of a modification to one dimension (or tolerance) on other dimensions.

Once in the hands of the users (not only the pilot users and the experts in chains of dimensions but also the ordinary designers), OI3C becomes, above all, a dimensions manager, which is much more than a "mathematical-graphical solver-translator" coupled with Pro/ENGI-NEER. Numerous observations and discussions with the pilot users revealed that the most delicate and the most boring task from their point of view is indeed the technical management of data: they focus their attention on it, and from there their expectations are expressed and possible resistance brought to light.

Finally, OI3C seems to have no definite frontiers. The solver (calculation tool) is supposed to be at the very core of the tool but is in fact only one feature among others. The interface with the other CAD tools actually attracts more attention. In the course of practice, the technical administrator, designed to be a useful plug-in, is transformed into a central feature, helping to tackle the problems encountered in the daily lives of the designers.

Conclusion

OI3C is hybrid and designed for multiple uses. It is an instrument of technical coordination, standardization, and promotion of a new mode of design. It is no longer a black box: the dynamics of validation gradually opened it. Starting out life as the answer to a need expressed on a specifications sheet, it grows into a tool that can actually serve a purpose. However, in 1997, it is still not up and running at the industrial sites. Nevertheless, in view of the pressing demands of the departments involved, the curiosity of the other divisions, and the enthusiasm of certain pilots users, everything seems to point to its imminent implementation. If it does become operational, it may become a means, among others, of strengthening the new identity of Green and promoting a new corporate culture.

We have noted to what extent the tool was built by different actors, including the users themselves. This co-construction has not actually come to an end. Even in the near future, when the "industrial" 1.0 version is installed on several dozen workstations, it will probably continue to be transformed in the hands of the users, and new versions will be required.

Operational Summary

1. A lot depends on a tool. It is therefore worth studying what is at stake. In this chapter we looked especially at the identity and unity of a firm, its training policy and its design product philosophy.

2. A firm's identity is shaped by its work instruments. These partly reflect the construction of the firm's industrial identity.

3. To really understand the purpose of a tool, it is useful to describe the situation in which it is used, as well as the objects, texts, people, and discourse with which it is associated. It is a question of looking into the tool's various occurrences, which may be material, textual, or verbal. The observer must not sort occurrences a priori into meaningful and meaningless categories; he must first objectively record the links drawn between them by the actors or as they emerge in situation.

4. The instrument is at the core of a network. Describing the network is like opening a black box, which is what the instrument is. Even the inner core of an instrument, its most technical part, can be analyzed if the actors involved are tracked during the course of construction. Next, the thread of associations and exchanges between actors must be followed.

5. The history of the design and development of an instrument teaches us many things about its nature. First, the instrument comes to life in a specific context, following a series of events that dictate its future. It is the subject of tradeoffs between actors, and it is institutionalized before being transformed into an instrument or a code.

6. Technical activities can undergo the same sort of analysis as objects. Thus, validating is not just a cut-and-dried practice; it has to be observed and analyzed. Simply saying "validation" is by no means tantamount to understanding what goes on. Its meaning can only be understood by taking into account the actual practices to which the term 'validation' is linked.

7. There is permanent tension between contextualization and decontextualization while an instrument is being developed.   The validation process presents an opportunity to manage this tension.

8. The nature or the identity of an instrument is not given a priori.   Its nature is unveiled as we describe how it is built and how it is used. Thus, OI3C, which was supposed to be a design aid tool, also appears to be a coordination tool and an instrument for standardizing practices.

3

Social Complexity and the Role of the Object: Installing Household Waste Containers

Michel Bovy and Dominique Vinck

The first two chapters dealt with situations and problems related to design and innovation in the context of organized structures. Most of the people observed were professionals. However, in the course of his work, an engineer or a designer often meets and negotiates with other people, and especially with clients. In this chapter, we will observe an engineer in his dealings with society. As designer, his role is to gradually integrate an increasing amount of information concerning the problem as the project is materialized.

Of particular note is the manner in which an engineer attempts to predict the behavior of groups of people through the transformation of objects and their use (and meets with relative success). He discovers an unexpected plurality in the world of the household, which the designer believes to be a socially homogeneous group, a world adhering perfectly to a single model of behavior. He finds the world of the household much less predictable than he imagined. The introduction of a new object reveals the heterogeneity of society. As the action is applied, splits and divergences appear that alert the designer to the necessity of developing a deeper understanding of social complexities. He finds that society is composed of social groups with different objectives, identities, interests, and types of behavior. Furthermore, these differences do not necessarily exist prior to the action. Indeed, the action produces, rather than reveals, types of behavior and discrepancies that might have never existed otherwise. When put into contact with new objects, speech forms, and rules, people react and spontaneously change their identifying characteristics. The objects themselves take on unexpected characteristics. In the course of such a project (introducing a new object with a view to establishing a new type of behavior), countless discoveries are made and a multitude of transformations are observed. Its history is the study of objects and human groups and their respective identities.

In this chapter, we will examine a project for selective sorting of household waste. It is designed by an agricultural engineer concerned with producing high-quality compost. The project mobilizes and affects nearly all facets of society: the balance of political power, the composition of society, the strategies of its members, the objects they use daily, education, industrial activity, legislation on community action, etc. During the ongoing process of the project, the engineer comes to a deeper understanding of society as he is transforming it. The project he is responsible for, the way he handles it, and his own identity are affected, as is the manner in which members of the community view everyday, trivial gestures.

Our focus will be on a technical element of this project: the waste container. Upon first analysis, it appears to be a very simple object. The head of the project for selective waste collection sees it only as a means of reaching his own ends. In fact, as the project evolves and analyses are undertaken, the object is distorted, becomes inseparable from other elements, is displaced, and is articulated in a specific manner within a socio-technical network. It becomes an aggregate (a stabilized group of objects, rules, and human actions) that mediates (that is, instigates and transforms) human actions and projects. It acts as a translator between partially heterogeneous social worlds. It is a focal point for building new social links and explicit achievements in waste management. Therefore, the concept of the object as mediator will also be explored herein.

In the history of this project, the container serves to crystallize overall social relations more than any other element in the aggregate. The project unfolds as if other elements (the rationale behind composting, the principle of selective waste sorting, etc.) had never been called into question. Every time a new social component is discovered or a doubt arises concerning one of the elements, the adjustments that come into play between those involved are applied back to the container, and lead invariably to its transformation. In studying the role of super-mediator that the object assumes, this chapter reveals the excessive importance the acting agents place on it relative to other elements of the socio-technical network. It soon becomes evident that this relatively exclusive role of mediator is technically limited. Attempting to take the whole of social reality into account, using only one object to ensure its coherence, brings this limitation to the fore: the situation becomes unmanageable. Managing the whole of social reality is clearly not feasible from a technical standpoint. It is also interesting to note that social groups are capable of getting by on their own (as in the case of existing arrangements between the Scouts and camp managers) and contributing to the project while breaking free of imposed mediation.

The Institutional Infrastructure of the Project

In March 1996, an experiment in the selective collection of household waste in a semi-rural community was launched. It provided an opportunity to reflect on the concept of material objects involved in community actions and their place in management. The objectives are to describe the infrastructures involved in this experiment and to understand the stabilizing role that material objects can have on people participating in such a project, either as citizens, as companies, or as administrative authorities. We also hope to gain a deeper understanding of their role as mediators or translators between socially heterogeneous worlds. We became involved in this project as observers and investigators, sometimes as advisors (relying upon a broad vision of the action we were gradually initiating and a few theoretical references), and at times as teachers of the personnel of inter-community organizations (such as the District Council). For two weeks we also filled the position of dispatchers at the two container sites managing the largest volume of waste and having the highest rate of user frequency.

The waste sorting project did not simply fall from the clear blue sky. It was in perfect keeping with the regional, national, and European contexts, and it coincided with the climate of public rights (especially those concerning community and inter-community legislation) that provides a framework for diverse initiatives, projects and their management. In the community studied, the District Council of the Department of Sanitation for Intercommunity Services approved (in September of 1993) an engineer's proposed project for joint waste management, with separate door-to-door collection of organic waste. The District Council adopted the project in May 1994 and the community of G was proposed as a pilot community. The District Council pledged to take on all costs inherent in the implementation of this new means of collection during its experimental stage. In January 1995, project associates visited the community of G and presented the project for the first time. We will now describe the events that ensued in a present-tense narrative.

The necessary experimental plot of land is attributed as a result of the deliberations of a decision-making authority (the District Council) and complies with existing legislative and politico-economic requirements. Nevertheless, this does not suffice to ensure the complete success of the operation. The involvement of the chosen community is essential to the implantation of the proposed project on a local level and to its operational success. The community must therefore cooperate, and its willingness to do so is not immediately acquired. This can potentially limit the District Council's innovative ambitions. As a matter of fact, the community initially puts up a certain amount of resistance and requests that the size of the project be justified. It stands up for its political autonomy and reminds the technocrats that technically innovative projects are always politically oriented.

In June 1995, the engineer who conceived the project planned its implementation in three stages:

1. The adoption of a separate collection scheme for all households by the pilot community of G. The District Council will be responsible for providing technical and/or organizational support staff.

2. A discussion with businesses and community groups in order to best meet their needs (for example, using additional containers and implementing waste collection more frequently). The goal of this discussion is to verify the degree to which container sites are accepted, to organize collection runs, and plan information flyers for the general public.

3. Finalizing complete specifications for the waste collection plan: organization and distribution of bins, collection runs, and launching separate collections for the entire population of the community in November 1995.

The project proposal provides for certain conditions to be negotiable, i.e., service rates and specifications. Once these are defined, the project can be considered complete, both in its general terms (which are in principle non-negotiable) and in its lesser details. All that then remains to be done is to set up the project and develop communications in order to gain the confidence and acceptance of all parties involved. From the start, the head of the project expects to meet with some resistance—this is only human nature—but he believes the problem can be resolved by means of a well-designed communications plan: information leaflets, an instructive agenda with a touch of humor, regular contacts with district councilors, and public meetings with the district's principal economic agents to answer any questions they may have. The most important thing is to find the right medium.

The Container as a Medium

Starting from the concept that, in theory, material objects make it easier to determine behavior types than simple negotiating, the head of the project conceives and mobilizes an array of objects: communication equipment, sheds for waste sorting, containers with separate compartments, etc. The project can thus be broken down into a list of object which, at a given place and time, are assumed to channel action so that the goals can be reached. The objects are determining factors both of relationships and of results. They convey and yet curtail the force of the decision maker's action in a predictable way.

By buying 2,000 containers with two compartments (one for organic waste, to be composted, and one for other waste, to be incinerated) and a technically adapted collection truck for emptying these containers, the District Council and the private contractor hope to reduce the level of human intervention needed to separate the two types of waste.[1] The partitions of the containers and those in the collecting truck provide a stable basis for the separation of elements previously sorted by citizens. Signs placed in front of the container site and leaflets in mailboxes are used in lieu of informing each person individually upon his visit to the recycling bins or each collection. When an action leads to a desired result, material elements supposedly diminish haphazard events, whereas human operators are esteemed less reliable. Action by way of material mediums is considered the most dependable means of ensuring the exercise of power and control over behavior. Objects determine interactions and lead to expected results more accurately. They are also more easily overridden, as they offer less resistance to change. They stabilize expected actions more easily than the alchemy of human mood changes. Material objects supposedly lead to achieving a goal conceived through technical rationality. A posteriori evaluation of real performances of the medium makes it possible to measure deviations and define the elements it acts upon most efficiently.

Several partners in the project share the concept of the object's serving to relay and amplify the action: the project head and the District Council, the container manufacturer, the contractor responsible for waste collection, municipal authorities, and users. For them, the object simply serves to achieve their ends in a fairly efficient manner. Yet, as it happens, the same object is the medium of diverse and actually quite divergent projects. Each project head claims the right to design, manufacture, install, tax, displace, and use the object as he sees fit. The president of the local business association, the head of the District Council, the campground owner, and the Dutch camper do not have the same expectations with regard to containers. Thus, the object and the action that it was supposed to accomplish reflect the diversity of the specific individuals concerned and their projects.

The Container as Mediator

The physical aspect of the object (the volume or shape of the container) is not limited to one and only one purpose—that for which its creator initially designed it. On the contrary, it is inextricably tied up in the complex play of purposes of the various agents involved. The message it is supposed to transmit and the action for which it has been designed become multiple upon contact with concrete beings (as opposed to abstract concepts such as "population"). The object and its physical aspect are surrounded by a multiplicity of intentions and actions that intermesh. A plurality of design, installation, and reception of the object now exists. Therefore, the object becomes a focal point and a working mediator for those who come in contact with it. Its performance cannot be reduced to an initial intention or to the a priori known characteristics of the medium. At the level of the object itself, unforeseen situations also appear which affect the interrelationships of the participating agents.

For example, installing containers with a capacity of 190 gallons near shopping centers created a controversy. The District Council, eager to optimize collection and minimize costs, considered several possibilities: regrouping the containers in storage blocks; leaving them in front of shops; or asking local merchants to keep the containers in their garages, taking them out only at collection times. But this reasoning does not take into account the material problem of decomposing waste, which has the annoying habit of causing unpleasant odors. Shop owners will not hesitate to complain. Bad odors are therefore taken into consideration in an anticipative fashion. They are thought of in relation to tourism and the aesthetics of the center city. The presence of wasps around an ice cream container will bother customers in an outdoor café. Greasy, smelly packets of fried potatoes could pile up in a bin next door to an exclusive restaurant.

Faced with this influx of unforeseen agents (luxury shop owners and fast food stands, smells, tourists, wasps), the District Council is obliged to act. The arguments it uses bring other entities into play: a filter, covers, locks, marketing, etc. The containers selected lack filters because the District Council's purchasing officer does not believe they are useful. In his opinion, filters are only a commercial argument of container manufacturers and have no technical foundation; the tight-fitting traditional covers should be sufficient to prevent odors from escaping and keep local merchants happy. This is how the socio-technical world of the "selective waste sorting" project gradually becomes peopled with malleable beings.

Questions that were essential to everyone revolved around the container. In the course of the project, small social phenomena such as NIMBY ("not in my back yard") developed, making it necessary for the District Council, the retailers, and the local environmental and tourism authorities to negotiate. For example, the District Council considers asking people to put their containers on only one side of the street, to avoid having the truck go down the street twice or to prevent accidents due to workers continually crossing the road; however, it is felt that one side of the street might feel burdened by this. The burgomaster (mayor), unwilling to sacrifice half of his potential constituency, opposes this facet of the project. The priest refuses to let the parking lot that is usually reserved for his congregation be turned into a container storage site for local merchants. He nevertheless admits that current environmental problems require active participation on an individual level, and that making individuals aware of their responsibility with regard to public welfare is an integral part of his moral teaching.

It proves impossible to find a compromise acceptable to the principal agents of the project that would permit placing the local merchants' containers in the heart of the city. A suggestion is made to put the containers outside the city center. From the District Council's point of view, this solution should not pose any problems, since it is technically feasible and economically acceptable: the retailers all have their own vehicles and can have access to a trailer. They can therefore transport their containers to the appointed place. The solution would have been definitively adopted had another difficulty not arisen: it is not practicable for each merchant to keep an eye on his container. Users could take advantage of the situation by filling a neighbor's container to avoid paying local taxes for the use of a second container. To circumvent this problem and adopt the solution of a refuse block outside the town, the head of the project proposes stamping each container with a number that can be matched with the user's name and address in the Commercial Register. However, human beings find it somewhat difficult to differentiate one container from another and associate a container with a person simply on the basis of a number. It is therefore decided that a paper label bearing the retailer's name will be attached to each container. Also, a flexible plastic film will be placed over the paper to ensure weather resistance. The written word is thereby added to the number, and the container becomes personalized. But within a week the rain begins to seep under the plastic film, degrading the paper and making the names illegible. The process of decomposition acting on the paper—highly desirable inside the container—becomes on the outside a detriment to the project of individualizing the object.

The container and its location translate the compromises that social partners gradually agree upon. Furthermore, unforeseen agents act upon containers: non-authorized users, waste in a state of putrefaction, tourists, rain, etc. To maintain stability and predictable behavior among all the active elements, so as to achieve the desired goals, the project head must mobilize an increasing number of elements (the containers, the Commercial Register, the retailers' vehicles, the disposal site, the plastic film, cooperation on the part of the population, etc.) and fit them into the scheme. As the project progresses efficiently toward its predicted performance, new agents and entities appear and interact, forming a peculiar configuration that transforms existing purposes and actions. It provides a focal point for mediation or a mediator.

However, a container that has become a mediating configuration (composed of a place, a link to the authority in charge of the Commercial Register, a written label, and predicted types of behavior for different categories of people) cannot determine all forms of behavior. Each new adjustment engenders reactions and other unforeseen entities appear. The container block for retailers rapidly becomes an illicit dump. Next to the legitimate container site, rubbish bags and scrap metal are illegitimately discarded. And some users deposit rubbish near the entrance to the container site (intended for removing recyclable waste: paper, metal, glass, plastic and bulky items) because they often arrive when the site is closed. Furthermore, excessive household waste has a counter-effect on the unoccupied space in the retailers' containers which the retailers are unable to survey.

The container is thus at the heart of various problems caused by the unforeseen diversity of actors. It then becomes an object of controversy and cunning stratagems. Gradually the container is modified in order to adjust to the various compromises, which in fact do not all have equal impacts. It acts as a mediator between those trying to implement the waste-sorting project and the project's effective performance.

The Container as Translator

The container—sometimes a medium (intermediary or commissioning[2]) and sometimes a mediator—is also a translator. It expresses in terms of other elements (especially physical ones) the intentions, plans, and values of the actors, as well as the history of the relationships they share.

Nevertheless, as is the case with any translation, the medium is treacherous: instead of simply conveying meaning and actions, it displaces and transposes them in a semi-unpredictable way.

A Material Translation: Can Social Action Be Kept under Lock and Key?

The task of the project's head is to create a method of waste sorting that will stabilize the behavior of the agents involved. How is he to accomplish this? Through the use of material objects, of course. The objects will supposedly channel human behavior via a series of physical constraints. In a manner of speaking, the end result will be determined by the way in which the material infrastructure was set up.

Containers are installed. Each person is supposed to throw his rubbish in his own container. However, the mere presence of the containers quickly causes unexpected types of behavior. In the summer of 1996, some tourists find them handy for disposing of remains from fast food meals or bags full of household rubbish. Secondary residents, in order to avoid taxes, do not register their addresses with the District Council. Consequently, they do not receive their special "Duobac" containers, and they are not able to meet the waste collection requirements. They use the containers of registered inhabitants to dispose of unsorted waste, which under normal circumstances has no place in this mode of collection.

Even neighbors who are registered sometimes behave in an undesirable fashion. Cocktail parties on pleasant summer evenings lead to more waste material than usual, and the containers of the nearest neighbors are quickly filled up. This lack of consideration then forces those neighbors to place their rubbish bags beside or on top of their own containers, and thus they are often faced with the inconvenience of having their waste refused at collection time or of having to clean up the mess left after animal scavenging. Moreover, the household containers, already relatively anonymous objects, become even more so when regrouped in communal refuse sites. Deterioration rapidly increases since their identicalness justifies confusion when they are being filled.

Community workers join the list of undesired users. By removing illegally dumped rubbish, they do the work of private collectors for free. As a result, they find they are now seen as additional staff for public waste collection. Unhappy with this and with the community's refusal to invest in a vehicle for their work, they put unauthorized waste in the nearest containers.

Still another problem is that of the exclusivity of the containers' use. A work group consisting of agents from the District Council and the pilot community re-analyzes the situation and concludes that there is a need to convince the population to respect—and to ensure that others respect—the private use of containers. The means of controlling this are partially assigned and translated to new objects: restricted openings, locks, and keys. In this way, the containers are transformed to better channel the behavior of the new agents that appear in the project land-scape. Iron bars are fastened to the containers to prevent neighbors and other intruders from depositing whole bags of rubbish in them. However, passing motorists stop at containers along the district's main thorough-fares and furtively wrench off the lids. In the center of town, the density of the population limits this vandalism. In town, however, road surfaces are not always flat, and containers sometimes roll out of place. A concrete base and metal rods are therefore installed, making it necessary to pro-vide a key to remove the containers at collection time. The container def-initely never serves as a simple vehicle for the selective sorting project; but constantly shifts its aims. Next, retailers—large producers of waste for incineration and a lucrative tax base for the community—demand that the District Council provide locks to ensure that each household and each business has exclusive use of its own container. But bags of rubbish are piled indiscriminately on and around the locked containers. The controversy shifts from the interior to the exterior of the container. Complaints regarding illicit use turn into reproaches concerning hygiene and the removal of unauthorized waste.

The same question incessantly comes back to haunt the engineer: How can human behavior be controlled? How can social action be kept under lock and key when objects supposed to act as channeling factors betray the project and modify actions? This is a matter of deep concern for the project heads. Before deciding on a plan of action, they explore the potential reactions of both human beings and material objects. Therefore, before locks are provided, negotiations between agents and with the material elements are undertaken. These negotiations trans-form the project while contributing toward creating a new version of it.

First negotiation: providing locks. This implies reopening the question of the relationships between agents. An exclusive agreement links the District Council to the private collector, who is in turn linked to the con-tainer supplier. With overlapping contracts of this kind, the community becomes entirely dependent upon one supplier. Neither the private appropriation of container use nor the installation of locks had been foreseen in these agreements. Moreover, as this is a pilot project, the col-lector did not think it would continue beyond the summer and only cheap containers were bought. They were not initially equipped with separate compartments,[3] or locks; a technician had to adapt them to the project by drilling holes in each of the plastic containers. The collection contractor opposes the installation of locks. If in the end he yields and accepts, it is due to pressure from the District Council, which agrees to finance the purchase of the locks. Nevertheless, in July, because most of the supplier's employees are on vacation, it takes several weeks to deliver a few dozen locks. (The District Council considers the extra expense of purchasing them elsewhere unjustified.) The containers will be equipped with locks, but their installation is delayed.

Second negotiation: the choice of locks. This is controversial. Three systems are available on the market: Y locks, tri-pans, and E locks. Y locks are supposedly more resistant but are more expensive to buy and install. Tri-pan locks are the least expensive and easiest to install. Furthermore, they are available immediately. But they can be easily opened with pliers, and they are less resistant to wear. Debates between the District Council and the collection company focus on whether container vandals are often equipped with pliers. E locks (the third solution) are easier to install and more economical in terms of labor required to render a container operational. This solution is, however, discarded, as E locks are judged too fragile when used for other purposes. After re-evaluation of the Y locks, the potential competence of vandals, the success of the pilot experiment, and the projected costs, durability becomes the deciding factor. Each negotiation results in a better and more firmly based understanding of the socio-technical world in which the project evolves and the new course it is taking. In this way, a durable lock inspires an irreversible vote of confidence in the success of the operation.

Third negotiation: positioning the lock. Having initially been placed on top of the lid of 60-gallon Duobac containers, they are later moved to the front and placed in a vertical position. The reason for doing this is to provide protection for the locks by placing them under the edge of the lids so as to prevent water from penetrating the mechanism and causing damage through rust or freezing. Once again, the project's future goals determine the choice of technique. The scales (translated by the positioning of the locks), vacillating between "a pilot project for the summer season" and "a pilot project to be adopted on a widespread basis," are tipped in favor of the latter.

The question of the lock does not always accurately translate the project. It actually betrays it at times, as can be observed with the "forest service containers" placed in wooded locations, picnic areas, campgrounds, and Scout camps. In response to requests from forestry engineers and game wardens, the District Council places containers with a capacity of 190 gallons along dirt trails regularly used by hikers and near recreational areas. The containers make it easier to store refuse in a closed, odor-free place, protected from rain, wasps, and small animals. They reduce the amount of work for forest rangers, allowing them more time to pursue activities other than rubbish collecting, such as making a list of Scout camps where containers can be placed and thereby creating revenues from local taxes. The container therefore acts as an articulation that liberates forest rangers from rubbish collection and allows them to focus on mapping existing Scout camps (several thousand in summer). The mapping of the camps enables the District Council to better control taxable agents and the flow of refuse. Being able to justify the origin and production of refuse makes it possible to reduce indiscriminate and illicit dumping and thereby justify the project's effectiveness to the Regional District Committee and the Inter-Regional District Council Committee.

An agreement is made with forest rangers. A work group is to examine how the behavior of the project's new agents (hikers and forest rangers) will be translated, while reducing organizational difficulties for the collector. Indeed, placing containers in forested areas is risky. It is feared that the material will be damaged. Fires can break out in containers partially filled with paper if smokers happen to throw their half-lit cigarette butts in them, even though these same smokers might sincerely believe that throwing their stubs in a plastic container, rather than on the dry grass, is safer. Forest rangers suggest putting sand in the bottom of the containers, but this would make it impossible to separate the waste from the sand.

More thought must be given to waste collection in forested areas. The dirt trails cause no problems during a dry summer. In autumn and winter, however, collection trucks will not be able to drive through muddy areas; neither their tires nor their motors are adapted to such conditions. Potholes in the road also cause problems: when a truck leans to one side while driving over an uneven road, its underside may scrape the pavement. A pickup truck is small enough to turn around at the end of the forest trail, but a collection truck must back up for a distance of approximately a mile. After only a week, the high cost for the District Council of the private collector's service becomes a serious problem. As a result, the District Council decides to limit the use of these containers. Furthermore, locals with cars leave bags of rubbish in these containers that should have been taken to the container site for sorting. The District

Council therefore begins to look into the possibility of fitting the containers with devices to limit their opening and to prevent illicit dumping.

If container lids are equipped with locks limiting their use to those with keys, then walkers and campers, for whom the containers were originally intended, will no longer be able to use them. The work group therefore discusses the possibility of limiting the lid opening. The openings has to be large enough to allow greasy papers, metal cans, and other consumer items typically used and discarded by hikers to be deposited, but remain small enough to prevent people from throwing in large bags that are indicative of resistance to the selective sorting project. It is decided that drilling an opening in the lid is the best solution. However, an opening will let rainwater in. A second hole with a plug in the container bottom is therefore necessary for draining. The precise location of the opening also becomes a subject of discussion, as the filling process must be optimized. If the opening is made on the left or right side of the lid, waste will quickly build up on only one side, the container will not be filled evenly, and half of the space will soon be obstructed. It was therefore decided to drill two holes, one on each side of the lid. Is this possible? To find out, the principle agents go to the workshop to familiarize themselves with the actual material object and glean some advice from the technician. The following day, with the aid of the appropriate tool, the container is rapidly transformed. However, the work of producing the revised models rapidly comes to a halt; complications having to do with drilling and with tools make it impracticable to continue the innovation.

Locks still haven't had their last word though. Containers are placed at the outset of the most frequented hiking paths for use only by forest rangers with a set of keys. This allows rangers to pick up litter left by disrespectful passersby and place it in the container. This solution is still not satisfactory; small bins along the paths are overflowing with rubbish next to containers with an ostensibly large capacity (190 gallons) which are inaccessible to hikers. For this reason, passersby don't think twice about leaving their rubbish on or around the big containers. Consequently, the forest rangers decide to hide the containers. Then, at collection time, as workers can't find the containers. Presuming that they have been stolen, they order new ones. After a new round of negotiations, it is decided that the containers will be placed in visible locations and that their lids will remain permanently unlocked.

All these setbacks to the project indisputably demonstrate that its translation into a materialized infrastructure geared to provide a channel for human behavior can succeed only when a great number of human and material agents and mediators are systematically integrated through a process of gradual adjustments. The translating medium is formed by interactions between all the mediators while being at the same time a constituent element of them. It contributes to establishing a stable framework for the action while neither guaranteeing automatic results nor remaining indeterminate and vague. Deviations can always occur, and adjustments often have to be made. Attempts to define fixed adjustments can in themselves create new dynamics. Time and again, the agents are incited to explore, comprehend, and organize the social and material world in which they move and on which they act. The material object is a mediating element in the project. Nevertheless, its action can be understood only by examining the dynamics of translation and adjustment that lead from one mediation to another.

Can Symbolic Translation Be Persuasive?

Instead of trusting objects to channel behavior, why not trust in the conscience and good intentions of the agents? If this were possible, good communication, informed citizenship, and adequate signs should suffice. A material infrastructure hammered out in endless negotiations would no longer be necessary.

Signs are therefore put up at the container site that enable citizens to place the right type of waste in the right container. Unfortunately, two symbols that resemble each other lead a good number of users to confuse the container for wood with that for bulky items. Moreover, the materiality of the symbol leads to other erroneous behavior. The presence of a trailer in front of the sign, or too many vehicles at the site, is enough to block it from view. In addition, the signs are not sufficient to induce proper sorting. Efficient sorting of plastic into colored and transparent polyethylene and high-density polyethylene depends not only on the symbol but also on the supervision and intervention of the employee at the site and the willingness of the population to comply. Just how powerful are symbols and words?

Even when provided with specific instructions, some citizens remain indifferent to the Duobac containers and to the requests to sort. The District Council therefore endeavors to make users understand the sorting process, its aims, and its procedures. An initial attempt at communication in the form a flyer describing the reasons for sorting, the responsible organizations, and the places in which each type of refuse is to be left meets with failure. Refuse remains for the most part unsorted: plastic bottles are found in the composting section, potato peelings mixed with bits of margarine in the section for incineration. In addition to the informational flyer, the project heads send a personalized reminder. A technician makes a tour of the household containers and distributes a note in mailboxes describing the quality of sorting observed, detailing objects considered undesirable with respect to the categories mentioned in the initial flyer. This action is linked to a legal decision that set a precedent by authorizing examination of the contents of a refuse bin after the waste is entrusted to collectors.

As the results of this intervention are still not satisfactory, other objects aiming at communication are mobilized. Now the idea is to guide the cognitive process of citizens by linking a particular type of waste to a compartment of the Duobac, Thus, the technician places stickers on container lids marked "for compost" in green with arrows pointing to the front section of the container and "for incineration" in red with arrows pointing to the back section. Will the sticker be sufficient to modify behavior? Will a certain type of weather-resistant glue that would prevent pranksters from switching the stickers around be necessary? Will it be necessary to provide instructions along with the sticker? How many other objects will be needed to make the first symbol less misleading? These are the questions that the promoters of the project are gradually led to ask themselves as they find that previous messages have failed to eliminate the problem of mixed refuse.

The Translation Is Still Incomplete

People and objects are never easily aligned by the actions of the project promoters. A new aspect of their being pops up just when the promoter's action is attempting to define their identifying characteristics and their peculiar and predictable behavior. In this same process, new facets of the project are discovered. Thus, project promoters have to re-analyze, re-evaluate, re-position, re-define, and re-organize the elements of their actions and everything related to them. Often, new elements crop up that will have to be linked to the whole to make it cohesive. The question of the project's limits then comes to the forefront. If all translations are necessarily incomplete, it also follows that a specific remainder eludes each new action. Unless a temporary limit is imposed, the process will be endless.

The project's aim is selective sorting. The task is to modify the behavior of citizens so that they sort waste efficiently at the beginning of the process. However, a troublesome hotel keeper suddenly appears on the scene. He finds it convenient to transport biodegradable elements from his kitchen to his father's wild boar farm. He removes the partition in the Duobac and diverts the container from its original function. He uses the container for rubbish collection well enough, but only for his own personal ends. The quality of his sorting is remarkable; the compost that could be obtained from it would certainly have been of good quality. Therefore, selective sorting is not an innovation as far as he is concerned. In his case, the visit of a communications commando to convince him of the positive effects of separating moist waste is superfluous. In his use of the object, the hotel keeper does in fact discover some of the designers' intentions: guaranteed cleanliness at the container site, easy handling (due to wheels), lightness of the container, and free distribution. However, his behavior interferes with the District Council's collection project. Its promoters notice the disappearance of one of their field agents (the container) and ask themselves how they are to guarantee that containers remain within their network. How will it be possible to demand the restitution of the transformed object? Can the private collector ask for the container to be restored to its original condition? How can use of the container be kept exclusively public? The hotel keeper is aligned with the project through the mediation of a new District Council ruling and a fine. Another solution would have been to allow the deviating element to escape and to espouse the hypothesis that a certain incompleteness is inevitable and not necessarily threatening to the project.

Translation: From Text to Field, or from Field to Text?

The selective sorting project proceeds through the mobilization of objects, symbols, and human agents. It also relies upon rules that give it legal leverage. These rules are elaborated and expressed in various official texts, notably in district regulations.

District regulations concerning the collection of household waste are the fruit of a collective process that brings together a variety of institutional agents, including the Municipal and District Association of Flanders and the legal services of the District Council and of the Regional Minister of the Environment. Based on diverse district regulations already in effect in Flanders, the group and its interpreters produce, after many revisions, a working document defining and articulating heterogeneous worlds. Until December of 1996, the definition of 'household' used in Flemish law was maintained unchanged. This definition included a person living alone; a group of people, with or without family ties, residing or cohabiting in the same lodging; institutions of public welfare; rest homes; schools; community halls; sports centers; youth associations; sports associations; shop owners, even those not residing in the district; secondary residents; and all vacationers, planned or unplanned. Upon first analysis, the population of the pilot project agrees with this definition. However, difficulties arise as soon as the field study to identify the social agents gets underway.

To establish a list of commercial activities in the area, numerous cross-references are needed: records of value-added tax, a list of businesses registered with the Chamber of Commerce, lists of places and people involved in undeclared seasonal work, lists of branch companies registered under the same legal entity, etc. Concrete identification of this nature is expensive, and the District Council finds it excessive. They believe that the selective sorting project should be able to do without such detailed socio-geography. Difficulties arise as soon as the question of imposing a tax is approached or containers are distributed. The legal definition of 'household' as per community regulations does not accurately translate the heterogeneity of waste production related to business activities. The regulations therefore must be revised, reorganized, and applied to both general and particular divisions. The organizational and fiscal network of waste collection can be controlled more effectively in this way. It is the link between human agents and their material production.

Certain definitions have been modified. French standards applied to the concept of unplanned vacationers better reflect the wide range of agents linked to seasonal activities and tourism. Manufacturing companies, small-scale producers, and service industries not located in the building where they exercise their activities are specifically mentioned. An additional category, "household and assimilated household waste products," is added. Similarly, "household waste and assimilated household waste" is precisely defined and limited to a maximum of 2 cubic meters per week. Beyond this limit, the District Council is not responsible for further waste removal; it is up to the householder. Thus, this law translates the collection truck's capacity, the collection time for excess amounts, and the District Council's desire to guarantee the legality of its waste removal service. The extent to which business (especially that related to tourism) is thriving in the district is reflected in the precise details given in the definitions. The community's legal responsibility is detailed in terms of weight, volume, collection schedules, and the nature of material objects collected.

Regulations also give a detailed definition of the relationships between humans and objects. The main purpose for this is to connect the container more closely with the selective sorting project. The regulations read as follows:

Collection of kitchen and combustible waste will be effected only when it is placed in special containers with double or single compartments and according to the following specifications:

• Containers are to be supplied by the District Council or by the company they designate and made available to entities falling into the category of households.

• Containers can be used by several households if they occupy an apartment building.

• Each container has an identification number

• Containers must at all times remain at the address they were originally assigned and delivered to.

• Containers cannot be disassembled, modified, displaced, transferred or taken away due to an eventual change of address or for any equivalent reason.

• Containers are the responsibility of the head of the household to whom they have been entrusted or his equivalent.

• Containers must be handled with care and used judiciously. The collection service should be immediately notified in case of damage, loss, or theft. . . .

It is forbidden to:

• Open containers placed at the roadside, empty their contents, take out and/or explore part of their contents, with the exception of qualified personnel acting in the course of their duty.

• Paint the outside of the containers or deliberately mark them in any way

• Leave containers along the roadside on days other than those indicated for collection, without special permission from the District Council.

• Place household waste beside or on the container. . . .

• Remove or have removed containers by anyone other than the collection service.

Following these regulations to the letter sometimes presents unexpected difficulties in the field. For this reason, one of them was rewritten. It was initially formulated as follows: "With the exception of prior written authorization from the mayor, it is forbidden to store waste with the intention of recycling it; this regulation overrules any other permits obtained and/or specific agreements reached." After some discussion, "waste" was replaced by "waste other than that produced through household activity." It is evidently not practicable to forbid households to store cardboard boxes or plastic packaging and bottles in their garage in order to dispose of them later at a container site. The objective of the original text was to dissuade people from storing large amounts of refuse on private lots, as scrap metal workers and automobile junk dealers do. Small and medium-size companies that stored cardboard packaging in their garage for several weeks in order to economize on transport were also targeted. After accumulating large amounts, they filled the containers at the site in a few minutes, preventing inhabitants from using this space and ultimately causing letters of complaint to flow into the mayor's office, or his challenger's. It was also necessary to prevent companies from leaving cardboard out in the rain for several weeks and then bringing it to the site wet and in a state of decomposition. This greatly reduces the quality of sorting and obliges the overseer to transfer it to the container for bulky items. The text therefore gives precise details on authorized and unauthorized behavior with regard to objects, waste, and containers. The text translates at one and the same time the social composition of local activities and the socio-technical system of selective sorting. Inversely, the sorting system supposedly translates the law. The rules, the project, the system of objects, the social composition, and the activities are all interrelated as a result of the incessant back-and-forth play of text and objects.

Translating Society in the Socio-Technical Network
Social agents cannot all be equally grasped and translated to the socio-technical network. The Scouts are a case in point.

In order to encourage youth camp leaders to be responsible, the District Council decides to request a deposit in exchange for the issuing of a key to access the container that has been put into place. In this manner, the public authorities can control the identity of waste producers and the cleanliness of the container site. Users are legally responsible for their container sites. This responsibility does not sufficiently guarantee the behavior of young people, in the eyes of the District Council. Therefore, rather than check the register to identify and then meet owners of fields rented to Scout associations, the project leader attempts to meet directly with those staying in tents for periods of less than 2 weeks.

It remains to be seen who in the camp will be considered responsible for the containers and for paying the deposit to receive the keys. On-site visits would be a solution, but would involve a lot of travel (there are several dozen camps) and be unpredictable (young people move around in the countryside or forest). It is not an easy task to get the message across to camp leaders. They are not identifiable through an official mailbox registered with the post office. They are not aware that they are required to declare their presence to community authorities upon arrival. Moreover, they fear having to pay taxes if they do so. The official list of campsites, notifying authorities of the location, estimated number of young people, and arrival dates, represents only a fraction of the total number of actual campsites. Without site visits, the camps cannot be identified. It is therefore decided to entrust the message regarding waste collection to the container itself. The technician employed by the community visits the area and attempts to meet with the young people concerned; he delivers the container and explains its use. The idea of having keys and deposits is discarded so that the container, left in place for use by successive groups of young people in the same area, can pass the message along. When the technician is able to meet with vacationers, he explains the selective sorting project and asks them to leave the appropriate refuse at the container site. In this way, the container left at the campsite can be used for two weeks. The messages are addressed to the cooks, who are more aware of problems related to rubbish removal, and to camp monitors, who often need to come up with daily themes for activities. Furthermore, instructive brochures concerning selective sorting are distributed. All that is now needed is to equip collection trucks with adequate tires for the rough trails leading to the campsites and detailed maps to locate the camps, based on the work previously done by the forestry engineers. In the course of the technician's visits, the official list of campsites is updated and the number of containers required increases by 40 percent. The question now arises as to whether it is appropriate to order new containers. Delays in delivery, the seasonal use of the containers, and the necessity of leaving them in place for the rest of the year limit the success of the project.

Despite considerable efforts to ensure collection, the results are only partially satisfactory. In the course of the collection process, the District Council discovers the existence of a close-knit social network between vacationers and decides to use this phenomenon as a relay for the collection action. Arrangements between youth campsites and campground managers have long been in effect. Scouts are permitted to bring their rubbish and throw their bags in with those of the campground, thereby taking advantage of the public collection service. They frequently come to use the pay showers and eat at the camp restaurant, which serves as partial compensation to the campground manager. Furthermore, campgrounds collect bags from undeclared neighboring sites rented out by the managers.

Once this network is discovered, the project head changes his collection strategy. That youth camps can utilize a 190-gallon container as an intermediary place to store waste also means that it can be used as a place of contact. Since the undeclared commercial activity is also linked to a number of 190-gallon containers, developing a stable foundation for these objects translates to the District Council's promise to remain discreet with regard to fiscal authorities and the campground manager's reciprocal commitment.

Conclusion

Paying close attention to the mediating capacity of material objects is not synonymous with resuscitating the credo of technocrats. Material mediation is not a self-sufficient means of determining behavior with respect to a given object. On the contrary, it is a complex overlapping that, upon analysis, proves to be a source of activity.

The preceding observation demonstrates the complexity of mediation through the object. It results in so many actions and adjustments that the object loses its status of object (a completely detachable physical entity) and is distorted or becomes a sort of hybrid. It is only under this condition that the object effectively and efficiently mediates the many actions that are gradually entrusted to it. It is collectivized, regulated, physically articulated, and reinforced. The project becomes more real but also more specific and more complex. Human behavior loses some of its theoretical freedom but gains the practical freedom to achieve at least one action.

Mediation also confuses causality. As the action unfolds, it is articulated and transformed. Factors of success and failure become more difficult to isolate. Conversely, the action becomes more tangible and opaque. Acting upon elements or variables of the project becomes increasingly difficult as it becomes more complexly and redundantly interwoven. At the beginning, many things were possible; however, as nothing had yet been done, no performances could be calculated. In the end, the network is so laden with countless mediations that it seems almost to be an autonomous and self-regulating system, propelled by its own inertia. It generates performances that can be identified, qualified, and quantified (in social or technical entities, in sorting rate, in cost, and so on) as long as it remains relatively stable. Nevertheless, it always depends entirely on the overlapping of multiple actions and local efforts of coordination, which are numerous and intermeshed yet still dissociable. All translations are incomplete and always generate a remainder or new element that can bring into question the infrastructures that have been so painstakingly put into place and secured. For the project head—an isolated actor—the project can be successful only if he is able to create this intricate overlapping, which requires encounters and negotiations with all interrelated elements (or, at the least, delegation of some of the work).

Delegation of work is central to deploying the action and securing its base. The project head believed that delegating to the object would be more than enough. A study of the actions of the various agents at work revealed that he was partially mistaken. Not only did many more agents than he had imagined come into play; in addition, the object itself betrayed his project. Time and again, it had been necessary to take new action and to redefine and re-delegate tasks. However, all things considered, the project head was only partially mistaken. Mediation through the object proved to be a stable enough basis for the action, but this was due mainly to the fact that the object had become a focal point for the mediation of multiple actions instead of being the medium of a single action. The container (transformed several times in the course of the project), the community regulations (rewritten several times), the manager of the campground, the mayor (who discovered new aspects of local policies), and pieces of public property are all examples of mediation or mediators.

Operational Summary

1. *The social world is not homogeneous.*   It cannot be described by a single model of behavior. It is made up of social groups with different objectives, identities, interests, and types of behavior.

2. *The nature of society is neither given nor intuitively cognizable.*   It must therefore be discovered through action that simultaneously reveals and transforms. Upon coming into contact with new objects, forms of speech, or rules, people react and spontaneously change their identifying characteristics. The objects themselves take on unexpected characteristics.

3. *Innovation implies articulating the different dimensions of social reality.* For example, it affects the balance of political power, the social composition, the strategies of its members, the objects they use daily, education, industrial activity, and legislation on community action. The project's success depends upon an overlapping of causalities at the local level. This requires encountering all interrelated elements and negotiating with them.

4. *The identity of the designer does not remain unchanged at the end of the design process.*   The designer progressively discovers more about society as he is in the process of transforming it and is thereby led to modify his actions and their results. The designer's identity is therefore affected.

5. Rarely can a technical object be reduced to the mobilization of a means to reach an end.   In the course of action and analysis, the object is distorted, displaced, and articulated to fit in with other aggregates (stabilized units of objects, rules, human actions).

6. An object mediates its designer's action.   In other words, it accomplishes the action and fulfills the intention that animates it while transforming it at the same time.

7. Delegation of work is central to deploying the action and securing its base.   The project head believed that delegating to the object would be more than enough. However, the object could only be partially controlled. If the object proves to be a stable enough basis for the action, this is due to the fact that it becomes a focal point for the mediation of multiple actions instead of being the medium of a single action. Moreover, the object is not the only mediator.

8. Exclusively technical mediation proves to be unmanageable.   Sometimes an object crystallizes the whole spectrum of social relationships. When this happens, the project unfolds as if other elements had never been questioned. With each new discovery, the object is thrown into question and transformed. It then becomes the overall mediator for the entire project. It takes on an excessive importance in relation to other objects in the socio-technical network. It soon becomes evident that this exclusively technical mediation is limited: taking the whole spectrum of social reality into account, using only one object to ensure its coherence, is unmanageable.

9. Social arrangements that escape technical mediation also contribute to the creation of an innovative project.

10. Mediation through the object depends upon actions and adjustments that change the object into a hybrid.   Thus, the object loses its status of object (that is, a physical and completely detachable entity). It becomes collectivized, regulated, and physically articulated. Simultaneously, the project becomes more real, more specific, and more complex.

11. Mediation confuses causalities.   As the action unfolds, success and failure factors become more difficult to isolate. Acting on certain isolated variables gradually becomes impossible as the project becomes more complexly and redundantly interwoven. The network is so laden with mediators that it seems almost like an autonomous self-regulating system. Its performances are contingent upon the overlapping of multiple interactions.

II

The Social Worlds and Cultures of Design

In part I we explored the socio-technical complexity of design and innovation practices. We supplied a framework and an analysis approach for getting to the bottom of and reporting on objects, instruments, and technical practices. In this way we were able to add new life to the empirical bases on which different understandings and models of industrial activity can be developed and translated into tools and methodological recommendations.

In part II we aim to qualify certain aspects of design activities we noted in part I by describing and qualifying the actors, their practices, the tools they use, the logic and reasoning behind their actions, and the way they see the entities they manipulate. We shall discover the tensions and complementarity between different schools of design and knowledge-production logic.

The Structural Engineer in the Design Office: A World, Its Objects, and Its Work Practices

Stéphane Mer

In this chapter I shall focus on the people who work in design office, striving to obtain a clearer picture of who they are and of the logic underpinning their approach. I shall suggest a series of concepts and a way of describing these people that takes into account the objects and work practices that make up their world.

A large number of people are involved all the way through the process of designing a new product: sales engineers, structural engineers, CAD operators, production engineers, machining operatives, etc. At first glance we may see them as people who contribute additional knowledge, specific to each one. However, they are more than just the "messengers" of certain forms of knowledge. They also act on the product and their action is based on objectives, priorities, and values that are specific to each category and its domain. The environment of each action is formed by a set of particular tools and objects that are used by individual actors in their design work. This set of tools and objects, which I shall call a *world*, is a consistent whole that structures the actions of these actors.

I shall focus particularly on the world of structural calculation. I shall begin by showing that each world is a consistent, structured whole. I shall then attempt to characterize these worlds, drawing on three concepts. It will be evident that these entities (which may be human beings, tools, or objects) belong to the same world, that they develop the same action-based logic, and that they share collective knowledge and a scale of values. However, all the entities in a given world are not identical. As will be seen, these differences are a source of controversy that gives the world a certain dynamic; furthermore, it is essential to take the form of these relations into account in order to be able to propose new design methodologies.

This presentation is based on observations made in a company that subcontracts for the aviation industry.

Organizational Framework

Bearings PLC is a medium-size company (roughly 2,000 people) that specializes in designing, manufacturing, and selling ball bearings for use in the automobile, railroad, and aircraft industries. Bearings fulfill a technical function and are used in numerous mechanical systems, guiding one part that is rotating in relation to another. We stayed in the company's aviation department. Before presenting the organization of the aviation department, I should explain two features of the industrial sector in which Bearings PLC operates.

In this sector, subcontracted systems were for a long time designed by the client. Subcontractors were simply responsible for manufacturing them. However, little by little, clients have asked their subcontractors to design products on the basis of specifications defining constraints and performance levels.[1] This has obliged suppliers to set up their own design departments. Air Bearings (a division of Bearings PLC that serves the aviation industry) based its department on the organization that already existed in Bearings' automotive division, where design work was already carried out.

The precision demanded in this sector and the quality of the materials used, in combination with the small production runs (50–200 bearings per year), mean that sales prices are much higher (10–100 times) than in the automotive sector. Cost is not a major concern in the design of these products. In addition, products of this type are difficult to obtain on account of their high precision. As a result, criteria related to high performance have taken priority over criteria related to ease of manufacturing,[2] to which designers do not pay much attention.

Within this framework, the organization at Air Bearings has the following characteristics:

• The production site, which is small (about 200 people), houses manufacturing, assembly, the production engineering department, and the sales department. The design office itself is located at Bearings's head office and comprises a design office,3 a research department, and a test laboratory. The two sites are about 10 kilometers apart. Despite being very small, the distance substantially reduces relations between the design office and the rest of the factory. In fact, most contacts occur at meetings (which are part of the formal design process) or when problems arise that have to be solved jointly. Most informal relations take the form of phone calls.

• The sites are also distanced by the fact that the design office does not belong to the same division as the rest of the factory. In hierarchical terms, it reports to the engineering division, which includes Bearings' other design offices, the research department, and the test laboratories. The factory is part of the industrial division, which comprises a sales department that does not report to the Sales Division. This site sees itself as a small company, because it has a full range of functional departments (except the design office).

Employees of the design office sum up this situation, which is uncomfortable both geographically and organizationally, as follows: "We fall between two stools." "We are assessed by people who do not know what we do and we work with people who have no hierarchical power over us." "We are treated as Air Bearings subcontractors."

The Design Process

Within the design office, the first aim in the design process is to respond to calls for tender. This is quickly followed by a brief study to define the product. This study specifies the main characteristics. A rough drawing is produced, and on the basis of this drawing a preliminary production cost is determined. This cost is then used to decide on a sales price and start negotiations. At the same time, a technical response is developed on the basis of a study using software that simulates the bearing's operation in situ. The technical and commercial negotiations that follow may last from 6 months to a year.

   If the company is awarded the contract—in other words, if the client places an order with Air Bearings for a set of prototypes (a dozen bearings)[4]—the second phase then starts. In this phase, the structural engineer carries out numerical simulations, using the SIM software package, to define the product's characteristics. On the basis of these characteristics, the CAD engineer produces the technical drawings. He prepares the drawings by making a compromise between the recommendations of the structural engineer and certain manufacturing considerations.[5] He creates two drawings: a component drawing and a manufacturing drawing. The component drawing is intended for the customer and will form the basis for relations between the two companies. Once the customer has validated this drawing, it becomes contractually binding. The manufacturing drawing is used by the production engineering department to prepare to manufacture the bearing. Next, the production engineers decide

on the manufacturing schedule and operation plans. Finally, the prototypes are manufactured.

The last two steps are important for the design process as they give rise
to changes, related to manufacturing difficulties.

At the end of the design process, a final component drawing is produced in consultation with the customer, who must validate any changes
made to the original definition. Subsequent production runs will be
based on this drawing.

Work Practices, Tools, and "Structural" Objects

Let us now turn to the structural engineer. As I have just explained, the
structural engineer intervenes twice in the design process. The first occasion is when the company answers the call to tender. The structural engineer prepares a technical response to the customer's needs. Later, once
the contract has been signed, he produces a more detailed definition of
the system. We shall focus mainly on the first step and the corresponding
work practices.

SIM Software: Built to Reflect and Guide the Structural Engineer's Work Practices

Customers define their needs in terms of functionality and performance.
They only see the bearing via the services it will render to the overall system being designed. This functional view of bearings is associated with
dimensional constraints. It has to fit into a specific space and meet certain technical requirements (surface finish, tolerances of outside surfaces, materials, etc.). These constraints are expressed in the call for
tender in the form of an "as-built drawing" showing the outside envelope
of the bearing and the neighboring parts. The envelope is the interface
between the subsystem and the rest of the system. Any changes to the
envelope imply changes to the whole system.

Air Bearings' work involves defining the "guts" of the product—in
other words, what enables it to fulfill its function. The task of the structural engineer is to characterize this function, specifying the levels of performance to be achieved. To this end, he translates the customer's
functional demands into criteria on the basis of which he can judge the
bearing's performance.

However, these criteria reflect the work practices and tools that are
used. The assessment criteria are in fact the same ones as in the results of
the simulation calculations. Clearly the simulation software being used—

SIM—plays an important role in the preparation of the technical response by the structural engineer. It simulates the operation of several bearings mounted on a shaft. It also takes into account the environment in which the bearing is to be used (shaft deformation, stress, materials, etc.). In addition, calculation time is short enough to allow for a process of trial and error in the definition of a bearing capable of meeting the customer's demands. It would therefore be useful to look for a moment at the development history of SIM.

SIM was written within Bearings PLC and was based on a general theory of how bearings work, enhanced by the know-how of the various engineers. Originally developed by a team of three engineers, it is currently maintained by two people. Changes are based on comments by structural engineers. The software is consistent with the work practices of the structural engineers for two reasons: they were responsible for its initial development, in line with their work practices, and they contribute to improving it, adjusting it to suit the use they actually make of it.

The SIM software formalizes the knowledge and the work practices of the structural engineers. It plays a central part in their day-to-day work practices. It reflects the way they look at the bearing. Let us now take a closer look at the software itself.

SIM simulates the operation of several bearings (up to 20) mounted on a shaft. It takes into account the flexibility of the shaft and the outside forces exerted on it. In other words, it takes into account the operating environment of the bearings. In fact, simulation consists of a succession of calculations of shaft equilibrium in several positions, taking into account the characteristics of the bearings mounted on the shaft. These characteristics take the form of quantities and are consistent with the theoretical model that underpins the numerical simulation. The bearing is described by a list of figures obtained from the calculations and making up 20–30 pages of tabulated figures. For the structural engineer, these tables describe the bearing's operation. If he concludes that its operation does not satisfy the customer's demands, he changes some of the characteristics and starts another calculation. This process of trial and error is complete when he decides that the bearing meets the customer's requirements.

The structural engineers' view of bearings can be described on the basis of these work practices and this software. Each bearing is a technical system fully defined by a list of parameters. Its operation can be determined from the results of calculations, as it is modeled using the appropriate scientific knowledge. Once the structural engineer has defined all the bearing's parameters, he considers its design finished. All that is required is to translate the parameters into technical drawings so that the manufacturing department can then produce the bearing.

Observers not making allowance for this way of seeing the bearing will not be able to fully understand the actions of the structural engineer.

The Structural Engineer's Activities: Theoretical Modeling of the Bearing

Using the SIM software to define a bearing that meets the customer's demands is only one of the tasks of the structural engineer. Others concern the development of new products. In the course of the second stage of the design process he defines the bearing more precisely, but his attitude to it does not change. He still sees it as a technical object that has been perfectly modeled and defined using a finite list of parameters. He continues to use the SIM software, as well as other computer applications that enable him, after numerous adjustments, to describe the characteristics of the bearing in detail. These applications were also developed in house. They focus on specific features of the bearing, defining internal clearance, path curvature, etc. The aim is to design the best possible bearing in line with the customer's demand—in other words, a bearing that, when its operation is modeled, conforms to the requirements of the customer.

Another aspect of the structural engineer's work is not directly linked to design. For instance, he may take part in research projects, and he may contribute to improving the SIM software or other computer applications. The common denominator in all such tasks is that they contribute to developing knowledge related to bearings and their operation. What is more, this work is theoretical, involving theoretical constructions that may be used to model the bearing. If a fault appears in an existing bearing (during tests, for instance), it can be rapidly conceptualized thanks to the engineer's theoretical understanding of the problem. The objective is to understand why numerical simulation did not foresee this defect and then integrate the new knowledge into the various simulation programs. Similarly, new materials are studied to determine whether their performance is suitable for use in bearings, but also to model their behavior using the various software simulation tools.

Thus, it is apparent that most of the structural engineer's work is devoted to representing bearings theoretically and modeling their operation. With the progress in calculation code, modeling work has become essentially digital.

The Structural Engineer's Objects: Theoretical Objects

The first thing you notice on entering a structural engineer's office is that there are no geometrical drawings, which are usually omnipresent in design offices. On the contrary, the office is littered with sheets of paper containing seemingly endless data tables, graphs, and equations. There are not many drawings of any kind, and those that are present are neatly folded—proof that they are rarely consulted.

The objects created and used by the structural engineer are mainly folders full of calculations and computer printouts of simulation results from various software packages or test reports. Anyone not in the know would never suspect that all these documents refer to bearings. They contain innumerable symbols—α, z, d_B, K, G, H, d_P, d_{min}, P_{avg}, etc.—that are quite impenetrable until the visitor has determined their local definition. For example, d_P represents the working diameter, and d_B the ball diameter. These symbols refer to conventions, some of which are common to mechanics in general but many of which are specific to structural engineers. In other words, structural engineers have their own way of representing bearings, in keeping with their work. This method enables them to define the ball diameter, a curvature value, or a working diameter, or to choose an oil. It also conforms to the simulation software used.[6]

As we have just seen, the tools (software), objects (tables full of figures, symbols, equations, etc.), and work practices of the structural engineer form a consistent whole that is extremely interactive. The various elements define and mutually influence one another. They are based on a corpus of knowledge and conventions. In addition, their main concern is to increase theoretical understanding of bearings so that this can then be applied to the design of new bearings.

Judgments Made by the Structural Engineer

In the course of their work, structural engineers make a large number of value judgments. These may be divided into two categories: judgments that concern the tools, objects, and actions of other actors, and judgments that concern products being designed.

A structural engineer may, for instance, judge that the computer application that defines the bearing's internal play is no good. He has noticed that it produces erroneous results, or that it cannot be used to specify a clearance correctly. He may well suggest that the application has not been properly maintained and that it does not integrate the most recent knowledge.

On the other hand, the results produced by the SIM software are not questioned very much. It is thought to be the "best" simulation tool.

"That's what makes us stand out from our competitors," the engineers say. Similarly, the CAD software used by the draftsmen has a fairly low status.[7] Structural engineers never use it, nor do they know how to. They are not even interested in being trained to use it. This tool simply does not belong to their working environment.

And it is not unusual to hear judgments on other people. For example, when I mentioned an engineer who had left the department, the structural engineer described him as "very good," quite simply because he had contributed to developing the SIM software—all the proof that was required of the quality of his theoretical understanding of how bearings work. However, in the course of the subsequent discussion I learned that he had problems with his customers, and yet one of the key tasks in his job was to ensure that bearings corresponded to the customer's demands.

This shows that judgments focus on theoretical considerations. A practical tool, such as CAD software, has a lower ranking than the SIM software, which reflects the theoretical knowledge of the structural engineers. These judgments have little connection with the know-how and functions of the structural engineer as defined in the formal corporate organization. In the latter case, structural engineers must be on good terms with customers. Despite this, fellow workers may think an engineer is "very good" even if he has difficulty relating to customers. On the other hand, a structural engineer who is on very good terms with his customers but has only limited theoretical knowledge is not thought to be "very good."

Judgments in the second category concern products being designed. These products are based on the results of numerical simulation. A "good" bearing is a bearing whose simulated operation meets the customer's requirements. No importance is attached to manufacturing cost or feasibility. Choices and decisions are based exclusively on calculations and simulations. The latter show how a bearing will operate and take into account the way it will be used. However this assessment is exclusively technical, whereas there are other factors in the life cycle of a product— marketing, manufacture, maintenance, etc.

The judgments made by structural engineers all refer to a "scale of values," in the sense given by Boltanski and Thévenot (1987). This scale is based on ideas of what is good or bad, or what has a high or a low ranking. In the design office, the scale is closely linked to a theoretical understanding of bearings. A tool has a "high" ranking if it draws on theoretical models of bearing operation, but only a "low" ranking if it draws on other knowledge. Similarly, bearings are assessed entirely on the basis of their theoretical operation. We may conclude that there is a close link between the scale of values, which provides a basis for judgments, and the actions taken by structural engineers.

The World of the Structural Engineer

The above descriptions show that there is a structural engineers' "world" (Becker 1982). We shall define this notion of a world as a whole consisting of tools, objects, and human actors developing the same *action-based logic*, governed by the same *scale of values* and sharing *collective knowledge.*

In the world of the structural engineer, all the entities concerned contribute to a type of action relating to the theory of bearings. We could analyze this in terms of objectives, in terms of purposive-rational action, but this would not cover everything. There is also a value content. Individuals determine their actions as a function of the subjective meaning they attach to a particular gesture.[8] A structural engineer will perform simulation calculations even for a simple bearing, for which he could define the characteristics using his know-how alone. Acting in this way gives greater value to his action and design work. The term "action-based logic" (Karpik 1972) means that there is a certain continuity between all the actions of an actor, a sort of constant or thread. What is more, this notion avoids the need to dissociate the framework of action (objectives, constraints, value) from the action itself. The structural engineer makes judgments based on a scale of values closely linked to the action-based logic guiding his work practices.

The work practices of structural engineers are also based on knowledge and conventions that are specific to them. This corpus comprises theories of the mechanical operation of bearings and particular know-how related to their action. As a result, it is difficult, without a long learning process, to understand and use the tools and objects that structural engineers manipulate. On the other hand, this knowledge is shared by all structural engineers and thus constitutes "collective knowledge."

The three dimensions characteristic of the structural engineer's world—action-based logic, scale of values, and collective knowledge—are interconnected. This knowledge is built up and put to use in the course of action so that the action can be completed; it is in phase with the action-based logic. What is more, the action-based logic is influenced by new knowledge. Digital mock-ups came into general use at the same time as progress in computing, changing the work practices of engineers.[9] On the other hand, the boom in the use of simulation software is due to the fact that it corresponds to the action-based logic of structural engineers.

The second point that this conceptual aside highlights concerns the relations between structural engineers and other design actors within the company. Cooperation among the various actors is not just a cognitive issue; it also concerns different languages and product knowledge. Allowance must also be made for the other dimensions of their action and the scale of values on which actors base their assessments of design decisions.

The World of Calculation: Stability and Instability

The description above suggests that structural engineers are all identical. In fact they vary a great deal. There is no lack of controversy in the world of structural engineers, and it is constantly changing.

Several Ways of Being a Structural Engineer

There are three structural engineers in the design office at Air Bearings, each with his own work practices.

First, there is Jean, who is in charge of bearings for machine tools.[10] This particular market stands out for the high number of different orders placed. Jean is involved in a large number of projects at the same time. It is not unusual to see him complete a study in one day, for the ability to react quickly is essential in his job. He performs only a few simulations for each bearing. His empirical knowledge is such that he can design a suitable bearing without doing a great many tests. This is not to say that the SIM software package does not play an important role in his work. On the contrary, he spends a large part of each day in front of the screen, defining and modifying bearings and starting new simulations. However, he remains slightly wary of the results. He will sometimes change results because he thinks they are wrong. In fact he does not trust the software completely: "You have to be careful with software that gives you cut-and-dried results." Relationships are also an important consideration for Jean: "What I like about this work is the contact with customers, the arguments, the pleasure you get from solving a tricky problem." He is not particularly interested in research. "I don't have the time for that!"

The second structural engineer, Pierre, is in charge of bearings for aircraft engines. He has only a few customers, and he does not often need to develop several products at the same time. On the other hand, the stakes are high when he designs a new bearing. It is important to win the contract. It is crucial to design the best bearing so that Air Bearings will get the job. Pierre performs a large number of numerical and other calculations to optimize weight without impairing performance.[11] He can draw on vast theoretical knowledge of bearings and their operation. He aims to be extremely rational in his design work. For the trial-and-error process with the SIM software, he always proceeds in the same way, which enables him "to look at all the potential solutions." He takes part in numerous research projects, some in house and some with outside partners. Large numbers of new motors are not being developed all the time (only one was under development during our period of observation). In addition, Pierre is keen to develop his knowledge of bearings: "It is essential that we stay on the ball, to keep up with progress in research."

The third structural engineer, Jean-Pierre, is in charge of bearings for transmission gearboxes.[12] These mechanisms contain a large number of bearings (as many as 40, 30 of which may be different). When a new contract is being prepared, he has to design a large number of bearings at the same time. For the preliminary study, which defines the technical response to the tender, he uses the SIM software to design, as rapidly as possible, the bearings that will be manufactured by Air Bearings.[13] During this time he uses the SIM software a great deal. However, this sort of situation does not arise often—on average, only once every year or two. In the meantime, he supervises the various bearings for which major contracts have been awarded. He also takes part in research projects, but he devotes less energy to them than Pierre. He says he likes the alternating periods of stress and relative calm.

These brief portraits of three structural engineers show that they are far from being identical. They work in different ways. For one of them, modeling and numerical simulation is a way of speeding up the design of bearings. For another, the aim is to develop sophisticated theoretical knowledge so as to be able to design the best bearings. They do not have the same ranking. On the structural scale of values, Jean is not a "high-ranking" engineer, even if everyone agrees about his gift for rapid design and good relations with customers. Pierre, on the other hand, ranks high on account of his vast theoretical understanding of how bearings work.

However, despite the differences described above, these engineers have many things in common. The SIM software plays a central role in their daily work. They have exactly the same view concerning bearings: they are technical systems that can be completely defined by a list of parameters, and their operation can be modeled. Their work practices are based on theoretical modeling of bearings, even though they use different approaches.

The Dynamics of the World: Controversies and Relations

These three portraits also show that the world of structural engineers is full of controversy. Each engineer thinks his work practices are the right ones. "There is no point in designing technically perfect bearings if we do not get the contract because we are too expensive or it does not meet the customer's requirements," says Jean. "But Air Bearings' strong point, and the basis of its image, is the technical quality of the bearings it develops." "It is essential that we retain this advantage over our competitors," says Pierre. The same split was evident when we asked people in the design office why they work in the head office whereas everyone else at Air Bearings works at the factory. Pierre emphasized the proximity of the research center, the test department, and the calculation resources available at this site. Jean, on the other hand, would be happier at the factory, for this would allow closer links with the other people involved in developing bearings (those in the sales, production engineering, and quality departments).

Controversy is a driving force for change in this world. As compromises are reached, the world evolves. If the design office were to move to the factory, this would create new relational dynamics with the other departments. Theoretical considerations could come to play a smaller part, and the structural engineers would make allowance for matters related to manufacturing and cost. Indeed, the current status quo as to the importance of theoretical knowledge could be consolidated by strong links between the results of calculations and the drawing of bearings, with partial automation of drawing operations based on the characteristics defined by the structural engineers.

These examples show that the dynamics of the structural engineers' world are also fed by the relations it develops with the company's other worlds.[14] The questions raised by the other worlds provoke controversies or add to existing ones. For instance, the problem of not being able to manufacture bearings at a low enough cost to sell them casts doubt on the work practices of the structural engineers (particularly with the economic climate currently prevailing in the aeronautics industry, where costs are increasingly important).

Operational Summary

1. The knowledge moving back and forth in the design office cannot be disassociated from the actors conveying it. The actors are not just messengers of the knowledge they convey; their action-based logic, their priorities, and their values give meaning to this knowledge. The knowledge they convey cannot be disassociated from the actors conveying it.

2. An analysis of design work practices reveals various sets of people, tools and objects with various points in common: action-based logic, shared knowledge, and a scale of values as a basis for judgments. We have called them "worlds." The differences between the actors belonging to a particular world are slight in comparison with their similarities. An analysis of the relations between actors may thus be organized in terms of diversity of knowledge, action, and judgment.

3. The "worlds" are consistent, structured wholes, but they also structure the actions of the actors in them. The actors rely on these worlds to direct their design work. The tools and the work practices, within a given world, tend to be consistent with one another. The three dimensions noted in the analysis interact a great deal, forming consistent wholes.

4. The perception of a product and of its design process is related to the world to which the actor belongs. The criteria for good design or the decision that design work is complete, for example, depend on the world to which the actor belongs. His actions cannot be understood without taking this into account.

5. A world-oriented approach liberates observers from an analysis based on the constraints of the formal organization and on relations between departments. Several worlds may well coexist within a single department. Realizing this enables observers to look at design work while taking tools and organizations into account. This approach is essential to define design aids suited to the new types of organization found in this activity.

6. Numerous conventions are at work, forming a framework for the work of actors. Some of these conventions are well known and widely taught. Others are specific to particular design offices and their histories. Local conventions are usually consistent with their action.

7. Controversy is lively in the social worlds of design, giving them temporal dynamics. Actors makes judgments about products, methods, and other actors. The differences between them are a source of controversy as to what subsequent changes are suitable.

5

Contrasting Design Cultures: Designing Dies for Drawing Aluminum

Nathalie Ravaille and Dominique Vinck

Designers, in the course of their work, must apply existing methods, technical standards, and scientific knowledge. These differ from one design office to another according to the products being created, the needs specified by clients, and production constraints. However, these factors cannot in themselves explain the uniqueness of each designer's work. Above and beyond existing techniques, designers acquire skills, which are sometimes difficult to pinpoint. This proficiency develops gradually as the designer gains in personal experience and as the company for which he works evolves collectively. The words often used to describe these specific characteristics are rather vague: 'tradition', 'style', 'culture', 'paradigm'.

This chapter aims to examine the workings of the conceptual and theoretically intangible world of designers, a world that affects the content of their work. A comparison between two companies will help to bring out the unique qualities of each. We will pay specific attention to intermediary objects, not as determining factors of the design procedure but as indicators of the manner in which problems are implicitly tackled and resolved.

This study on which we report here not only examined intermediary objects produced and utilized by its participants; it also examined various objects created for experimental purposes. Using the technique of participant observation, various tools were created and presented to subjects. The use, non-use, or diverted use of the objects made it possible to test certain hypotheses—based on observation and previous conversations—concerning the implicit aspects of these designers' work. Introducing new objects is thus an indication of tacit dimensions involved in designing.

The idea of a design culture was not used as a working hypothesis at the beginning of the study. It emerged slowly along with the discovery of implicit elements, unrelated to tools, methods, and intermediary objects, which appeared to structure the designers' work. This chapter is an attempt to describe these implicit elements in relation to the observation of objects and actions. We will show that from one company to another, within the same industrial sector, working on the same type of product in the same country, cultures of design can vary widely. We will be describing two cultures, one based on calculations and rules and the other on overall perception of shapes.

An Unusual Design Process

The design process we will examine here is not generally representative of design work. It occurs in a particular technical and economical context, the characteristics of which we will attempt to describe in detail. The unusual nature of the process is, however, valuable in that it draws attention to the implicit culture that is specific to each design office. Therefore, we will first give a contextual overview of this particular design process.

Our study focuses on an industry that produces pieces of shaped aluminum such as window frames, ladder steps, bicycle wheels, and boat masts (figure 1). These extrusions are manufactured by passing a mass of aluminum through an opening that in some cases includes branches whose shapes modify the end result (the quality of the section or the speed of the drawing, for example). The opening is punched out of a steel disc that is referred to as a *tool* or a *die plate*. The tool must be strong enough to resist the pressure applied to the aluminum to force it through the opening, or it will break. An industrial firm that draws aluminum in order to produce sections is referred to as a *drawer*. The worker who operates the press is known as an *adjuster*. A drawer obtains dies from a *diemaker*, who designs and produces the dies according to the extruded sections the drawer wishes to achieve. The design process we will study in this chapter is that of the diemaker.

Designing in an Industrial Context

The goal of a designer is to produce a steel tool (the die) having one or several openings that might be partially obstructed by branches (for example, when a hollow extrusion such as a tube is desired). The designer must determine the number and shape of the openings and the branches.

A design office in this type of industry, typically comprising ten designers and five programmers, deals with a much greater volume of design

Bloom of Aluminum to be Drawn

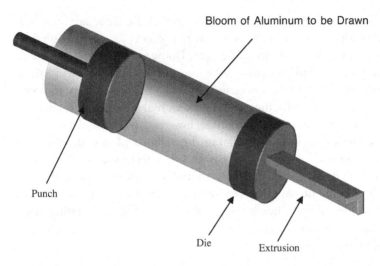

Punch

Die   Extrusion

Figure 1
A drawing press.

work than design offices in other industries involving mechanical engineering skills. Approximately eight dies are designed per day. The average design time is very short. In general, a designer spends between two and eight hours on each die. This time frame is radically different from what is typical in, say, the automobile industry, the end product of which is so complex that a great many parts, designers, and skills are needed and several years are required to complete the design. Here, on the contrary, the design process is very rapid. This is the first characteristic of the process we studied.

Of the eight dies designed daily, two are simply revised designs, or adaptations (made after observing results obtained in the drawing process) to already existing dies; four of the eight are designs that closely resemble dies previously produced; the other two are genuinely new designs. Designers have very few references to rely on in designing completely new dies. They must therefore spend more time considering the various problems involved. Each week, three or four dies require truly creative design capacities. This calls for a high level of creativity. In other mechanical engineering industries, designers work essentially on improving existing series of products in technical fields that are already well developed. Often a designer's task involves revising the proportions of existing objects rather than creating new concepts. Creativity is more central to a diemaker's work.

Diemakers agree that economic constraints in the trade are an obstacle. The market price of a die is so low that diemakers are tempted to come up with quick and cheap designs. Doing otherwise would mean selling the dies at a higher price, which would call for convincing their clients (the drawers) of the necessity to do so. They would have to prove that using more expensive dies would present real advantages in terms of higher output. This is not easy to prove, however, as professionals in this field lack proper tools for calculating the viability of new designs and demonstrating their respective merits. In other technical fields, calculations can be more easily matched to the objects designed, thereby demonstrating the probable success of the design. The contrary is true in the case of diemakers. They have no effective tool for calculating. This lack is rather shocking to engineers and mechanics, who generally base their decisions on calculations.

Resources and Constraints of Design Work

Similarly, the process of designing dies is characterized by a lack of codified knowledge about the behavior of tools in a working situation. This can lead a designer or an engineer astray. He must define a die on the basis of experience and hope that it won't break during the drawing process. In other fields of mechanical engineering, to define an unbreakable part, the engineer is accustomed to defining the stress that it can withstand and only later describing its characteristics. However, little or nothing is known about the stress that dies must withstand. An ordinary machine operator faced with this type of problem would apply maximum stress, thinking that if he over-dimensions the part he can be sure that it will withstand the stress. In the case of extrusion, if the tool is over-dimensioned, it modifies the flow of aluminum and increases the stress that the die must withstand. It is a vicious cycle. The die is over-dimensioned so that it will be more resistant, but at the same time the tool must undergo increased stress and might break. Also, when a tool breaks, the first reaction of designers is to consider a thicker, larger tool, so that it will be stronger and more resistant. Doing so, however, overloads the tool and increases rather than reduces the risk of its breaking. Paradoxically, a finer tool just might be more resistant. The real problem is that engineers simply don't have reliable instruments for calculating the solution to a given situation. This uncertainty and lack of knowledge with regard to the behavior of technical elements (dies and flow of aluminum in the press) is a second specific characteristic of this profession.

To date, there are no instruments for dimensioning dies adequately—
that is, no instruments that can predict, before testing a tool, whether it
is operational. Research engineers in this field are endeavoring to design
instruments for this purpose. Their principal approach is to establish theories that will shed light on influential factors and general tendencies
regarding relationships between certain parameters. This involves qualitative physics, the branch devoted to approximate theories. In any case,
adjusters working on presses as well as designers have a natural tendency
to reason in this way. They say to themselves "If this part is larger it will
have more influence on results, whereas if this element is larger the effect
will be insignificant." They apply principles of physics that define orders
of magnitude. Thus, professionals in the trade call upon an intricate web
of empirical knowledge; they know that if they act in a particular way,
results will attain a particular order of magnitude, whereas if they act in
another way, results will be greatly decreased. At this stage, the task of the
research engineer consists in formalizing his expertise, making its underlying principles clear and his approach tangible. This state of affairs may
be attributable to the relative youth of the drawing industry. Little knowledge has been gained about the extrusion of complex shapes, which goes
back only about 20 years.

As we have seen, designing dies requires a high level of creativity.
However, we know very little about this creative process—what prompts
the thinking process, the association of ideas, and the creativity. And yet,
designers are creating new ideas in design offices almost every day. Every
week technicians, not creative artists, go through the process of creation.
However, as they were schooled in mechanical design, they are in no way
prepared for this aspect of their work. Indeed, one of the firms studied
employs an artist. Its head of design, who formerly worked as a commercial artist, is well known for his remarkable craftsmanship and ingenuity.
In fact, his firm appreciates these creative abilities more highly than did
his former employer, which did not exploit his talent at all.

Design, Cost Control, and the Culture of Industry

Yet another unique characteristic of design work is related to industrial
organization. Diemakers (designers) are employed by specific manufacturing firms (drawers); this is true even when the firms belong to the same
industrial group. After the cost of raw materials and aluminum, expendi-
tures on die sinking are the highest item on the budget, coming even
before labor costs. Because of this, in an effort to reduce production costs,
drawers tend to concentrate on this budget item and attempt to find

cheaper means of making dies. They put pressure on diemakers, who in turn tend to spend less time designing each die. This holds true even when diemakers and drawers belong to the same industrial group. Yet overall cost analysis demonstrates the advantages of designing more productive dies (dies that double the drawing speed, for example) even if they are initially more expensive. Managers of the industrial groups concerned have recently been trying to get this message across. On the other hand, these same managers evaluate each manufacturing company individually on the savings made. The decisive factor in this contradictory policy is that the manufacturers are more concerned with their cost-cutting evaluations than with reputedly superior principles of rationalization.

The tendency of drawers to constantly reduce the cost of dies is all the more pronounced because they produce a large number of extrusions and therefore they need many dies. In fact, 80 percent of existing dies produce only 20 percent of the pieces extruded, which means that the quantity they extrude is very small. Significant increases in production could be obtained from 20 percent of dies. However, since increases in production are not immediately evident, drawers pursue their rationalization policies with regard to immediately tangible elements: the design time needed and the cost of buying the die. Currently, one of the most important challenges is to develop the ability to verbally express and to categorize dies instead of relying on intuition. This means specifically describing what is expected of each die, what each one can accomplish, and why.

Finally, uncertainty about the reasons dies react in certain ways (breakage, defects, etc.) leads adjusters that operate presses to be extremely careful. They avoid pushing machines to full capacity, especially if they are reprimanded when tools are broken. "The man on the press doesn't want to break a tool because he's the one that will get yelled at, so he doesn't push his luck." Since no one can really predict the behavior of dies, adjusters are very cautious and guarded with them. Even so, a new culture is emerging in these companies. To increase production, engineers and press foremen put pressure on adjusters to run machines at full capacity, despite the risk of breaking tools. All these elements come into play in the workaday world of die designers.

The Design Process

Now let us examine the overall process step by step. We will begin with a client placing an order with a drawer, then discuss how this affects the diemaker. Finally, we will follow the die back to the drawer.

From the Client to the Design Office

A client, for example a joiner in the construction business, needs a certain type of aluminum window frame. He makes a diagram of the extrusion he wants and contacts the drawer. Either the client asks for samples and the drawer has no idea of the quantity of pieces that will eventually be ordered, or the client specifies the size of his order immediately (two tons per month, for example). Orders can be quite large, as in the case of rungs for ladders, or they can be limited and irregular. Some joiners request different extrusions in small quantities for each building. In some cases, the drawer's technical sales representative can tell the client whether making the particular piece is possible, specify what quantities can be produced, and estimate delivery dates. Between the time an order is placed and its delivery date (which can be as little as four weeks later), the die must be designed, created, and tested, and either samples must be drawn or the order must be filled. In other cases, the sales representative cannot answer the client's questions and asks someone who is more competent in the factory, generally the chief adjuster. At this point, a discussion between the drawer and the client can lead to the requested piece being modified because one of its parameters (dimensions, tolerance, or shapes) is difficult to achieve. Economic questions would also be discussed. If the client is a reliable customer and the drawer does not want to lose him, he will do everything in his power to make the piece. If, on the other hand, the order comes from someone who might not be a lucrative client or might only have small orders to fill, the drawer will not go all out, but will simply inform him that he cannot make the particular piece. The client might also say that he has made inquiries with competitors who claim to be able to make his piece. The drawer might then reconsider the problem and attempt to find other solutions by consulting different diemakers. The drawer will ask the diemakers if they can design a die for the desired piece. Depending on the responses he obtains, he will decide whether this type of extrusion is possible. Therefore, according to the order and the client, either the sales representative knows whether or not the drawer will be able to fill the order without consulting the chief adjuster, or the opinion of the chief adjuster is a necessary factor. Most often, he orders the die from one diemaker and does not solicit bids from others. These decisions depend on the degree of complexity of the piece required. This degree of complexity is, however, only partially formulated. Certain pieces present difficulties that are not easily identifiable and cause enormous problems in design or drawing.

Once the client's diagram of the piece has been accepted, it is sent to the diemaker who has been chosen to do the work, along with the parameters and a few precise details concerning the manufacturing context: the press that will be used, the quantity that will be drawn, the diameter of the tool, the number of simultaneous flows (between one and twelve), etc. The diagram and these brief details make up the initial specifications for the design. Often they are faxed to the appropriate partners as rush orders. Therefore, some parameters are specified at the drawer's, usually by the chief adjuster.

From the Design Office to Manufacturing the Die

Upon receiving the extrusion diagram, a design manager examines it and decides whether it has any resemblance to existing pieces, or whether it will require a new design. Depending on the firm, the design manager will either send it to a specific diemaker or put it in the pile of "new designs." A designer then takes charge of processing the order. His way of working varies from one firm to another. Once he has finished his work, he makes a detailed diagram of the new die and sends it to a programmer, who writes direct digital control programs. The die is then ready to be manufactured. Thus, for one die, two to four hours are devoted to design, and nine to ten days are required to manufacture the die.

Programming and manufacturing of the die does not simply involve following the specifications defined in the designer's diagrams. Diagrams of the new die include a front view and cross-sections. However, the die has curved surfaces that are not entirely defined in the diagram. It might be compared to trying to define a person's face with simply a front view and a few cross-sections. Therefore, manufacturers must interpret certain characteristics of the die based on the diagrams.

Though they endeavor to remain faithful to the diagrams and their specified dimensions, they must nevertheless interpret the diagrams and try to imagine the shape of the die in the areas that are not represented in the diagram. In addition, since certain parts of the die are made by hand, their exact parameters cannot be defined. Even if these parts are machine made, results vary according to the machine used; moreover, the drawing process is very sensitive to the slightest variations in shape. A difference of a few tenths of a millimeter on certain dimensions can have a considerable effect on results. At other times, the specific dimension has no effect on results; designers don't always know why this is the case.

The outcome of this whole process is the creation of a die (that is to say, a steel tool) and the accessory diagrams resulting from the design process. Nevertheless, in the course of manufacturing this die, the necessary interpretations made, the particular machine used, and manual interventions mean that the actual die created sometimes differs from the die that was originally designed and represented on paper. The die is a concretization of the design that was not completely represented in the diagram. It is therefore a sort of translation, since the manufacturer was obliged to complete the design of the parts that were not defined, not to mention the fact that he might also have misunderstood the designer's implicit intentions.

From the Diemaker to the Drawer

The die is then sent to the drawer who ordered it. It goes directly to an adjuster at the factory. The adjuster, as is standard practice, checks certain measurements to make sure they coincide with the diagram. He touches and feels the die. He tries to run his hand along the path that the aluminum will take. He examines it from different angles. The designer cannot scrutinize the die in this way—first because it is a three-dimensional object and it is very difficult to have a spatial representation of it beforehand, second because systems of computer-aided design often are complex and take much longer to use in 3D mode than in 2D. To cut design time, diemakers therefore avoid working in 3D mode. But even if they did, viewing the die in 3D on the computer is not the same as examining the real die. The adjuster can see things on the steel object that few designers would be able to predict. Even if they could visualize them, their perception would be different from that of an adjuster touching it and identifying himself with the metal that will go through the die.

Furthermore, the designer never witnesses the drawing process. First of all, the only feedback he receives concerning the performance of his die are the comments made by the adjuster. Further, it is difficult for him to visualize in a precise manner, based only on verbal communications, the problem that the adjuster has discovered. Indeed, the adjuster finds it very difficult to describe what he sees. It seems, in fact, that to understand problems that arise in the drawing process one must be an experienced adjuster. And the designer always to some extent doubts what the adjuster says, for several reasons: for example, adjusters make apparently contradictory statements, and from one factory to another there are different traditions of adjusting. The process is so delicate that each adjuster applies personally acquired tricks of the trade. The designer therefore has a tendency to doubt the adjuster's observations, because he has trouble deciding which factors to take into consideration. And the adjuster can never accurately describe everything he sees and feels and all the factors he is considering.

Once the adjuster has examined the die, he sometimes modifies the form before even testing it. This irritates designers and research engineers. Little is known about the reasons for these modifications. The adjuster is hesitant to divulge his trade secrets, and even if he divulged them he would likely be unable to express them accurately. He proceeds differently depending on the die he is working with. Often he asks himself questions and carries out research. When the research engineer questions him, he doesn't always respond; he hasn't yet formulated an answer to the question himself. He is in the process of elaborating his answer, which is often the fruit of experience acquired with similar dies.

The adjuster then tests the die. He fits it onto the press and watches how the piece comes out. Multiple factors come into play at this point: whether the head of the piece is extruded upward or downward; whether the flows converge or diverge; how faithfully the head draws in relation to the rest of the extrusion; the smoothness of the piece; whether there are defects; verifying cuts at the beginning and at the end of the drawing as well as dimensions; measuring the length of the extrusion to make sure it conforms to specifications. Once the test is finished, the die is dipped in a sodium solution for cleaning. The adjuster measures it again to see if it has been modified in the process. From all these elements, he decides whether the piece corresponds to the client's order. If not, he modifies the die again (milling it, adding material to it, etc.) in order to adjust it differently. He then performs another test and re-adjusts the die until he obtains a sufficiently efficient extrusion that corresponds to the order. In the majority of cases, one, two, or three adjustments are necessary.

Adjustments might also be necessary during the drawing process itself. When this happens, the die is taken off the press and given to the adjuster who modifies it until it functions properly. With certain extrusions, the adjusters know that a die must be re-adjusted every three tons: at times the piece draws well, then it will be convex, then flat again, and finally concave. In fact the adjuster doesn't always know what causes these problems. For other types of extrusions, no adjusting is needed at all. Adjustments can also change drawing speeds by as much as a factor of three. Improving die performance during drawing can be achieved through an intervention in the design of the die.

Feedback and Communication between the Diemaker and the Drawer

At times, even after carrying out several adjustments, results are still not satisfactory. It is then necessary to have the diemaker redesign the die. In either case, the adjuster informs the designer of the modifications he has made. For example, if he has added material to the die by soldering, he will draw a diagram of his modification. The diagram is always simply a front view, a side view, or a cross-section. It is then the designer's turn to interpret the corrections and attempt to visualize what the adjuster has actually done and why.

If the adjuster sent a wax or clay mold instead of a drawing, it would be easier for the designer to understand. Problems of communication between the two stem more from the type of intermediary object used than from power struggles. Certain objects simplify communication. In this regard, mechanical drawing does not adequately express the complex information designers and adjusters need in order to make progress. One solution would be for designers to work on a regular basis (one month per year, for example) on the adjustment process. This would help them to develop the culture of the trade. It might also change the quality of future communication. Setting up this type of program sometimes conflicts with designers' contracts: they are hired to accomplish a task behind a computer, and employers cannot expect them to do work in adjustment. The problems are made even more complex by the fact that drawing and design are two distinct processes. In addition, in view of the cost-cutting constraints, designers' employers are not willing to pay them to do adjusting in one of their clients' companies. What advantage would there be for the drawer to allow a designer who knows nothing about the drawing process to work for him? With that kind of on-the-job training, both firms would suffer short-term losses.

Communication between designers and adjusters must go through the head designer—the person in charge of the most difficult designs. It is he who visits drawers and redesigns dies sent back from adjusters. This means that he is the person who listens to adjusters, tries to understand their experience working with former dies, and then re-transmits this information to the other designers. To a large extent, the context of discussions between designers and adjusters and the fundamental knowledge they draw upon are implicit. For example, the adjuster neglects to explain certain aspects of the problem because they seem obvious to him. One such aspect is the manufacturing context that guides the adjuster's decisions, but that he doesn't often think to explicitly mention. Also, in order for the designer to obtain certain crucial information, he must first be aware that the information exists and appreciate its importance; he must come up with and ask questions. This implies having some knowledge of the problems involved in the drawing process. If this is the case, the designer will try to obtain information.

In certain companies, adjusters are turning to designing dies themselves. They make a rough diagram of their ideas, and this makes their discussions with designers more fruitful. In one company, adjusters have higher qualifications than are usually required. They can therefore better understand the concepts underlying design work and even train other adjusters in design techniques.

Different Design Cultures and Practices

Let us now take a closer look at the activity itself. We will see that from one company to another, practices vary widely. We will demonstrate that these variations have to do with cultural differences and ingrained modes of thinking that are, to a large extent, implicit. Our focus will be on two companies, which we will call Famiform and Reglocalc to preserve their anonymity.

The two firms do not have requests for exactly the same types of extrusions. Famiform accepts orders for extrusions that are more difficult to manufacture. On the other hand, orders at Reglocalc are easier to manufacture. That being said, the complexity of the manufacturing process does not only have to do with the shape of the extrusions alone; tests are also performed to improve output and reliability. Aside from this, Famiform's research and development department is better staffed (with five researcher engineers) than Reglocalc's (one research engineer); this increases Famiform's opportunities for making progress. Finally, the culture of die sinking varies according to designers. Many of them have more than 10 years' experience in design work. At Reglocalc, those with more than 10 years' experience spent at least 5 of those years using computer applications for dimensioning. At Famiform, some designers have more than 20 years' experience. At Reglocalc, the oldest designer is 38 years old; at Famiform, the oldest designer is 50.

A Family Culture

Famiform has been in the design business since the start of the extrusion industry at the beginning of the twentieth century, and it has been operating on an industrial level since 1950. Drawing has evolved, and increasingly complex pieces have been extruded. The design department at Famiform dates back to the beginning of extrusion and has followed its evolution. As there was no system for calculating dimensions, designers had to develop different procedures. For a long time, dies were designed by adjusters, who manufactured each tool as they were designing it. After the drawer and diemaker trades split, a degree of historic continuity was preserved nonetheless. Working together, drawers and diemakers gradually built up a system of categorizing problems and solutions.

One designer, a former commercial artist who now works as design manager for the diemaker, developed the concept of "extrusion families." Without verbally expressing the characteristics of each family of dies and/or extrusions, he produced diagrams and families of diagrams. These simple diagrams give a synoptic perception of a set of family characteristics that cannot be easily and methodically expressed in words. (It is difficult to reduce each family to a few formal characteristics.) Not only would describing the characteristics of each family take a long time; the descriptions would invariably be incomplete, because a diagram coincides with an overall and complex perception acquired over long years of experience. A diagram also integrates results obtained from previously designed dies during the drawing process. If the die drew well, the designer works from the existing diagram and adapts the dimensions without using any system of calculation. "If it held last time, then I'll use it again and it should hold."

The design manager delegates work in the design office. He classifies clients' orders according to pre-defined die families. If an order does not fall into any of the known categories, he turns it over to the head designer. Since each designer has different skills, the design manager delegates work according to the abilities he believes each designer has. When one of the designers has finished his work, that designer checks the work schedule to find out which designs have been assigned to him. He takes the diagram and the specifications that go with it. He looks through the families of pre-defined dies to determine which family his assignment best corresponds to.

Above all, the designer tries to find which type of previously designed die is most like the new order. Having found a similar category of model, he attempts to class previously designed dies within the family. However, the notion of "similarity" is not very precise. The criterion for judging similarity between two shapes varies from one category of shape to another. At Famiform, this notion of similarity was defined and partially developed by constituting families of shapes.

The table of families of shapes lists serial numbers of existing extrusions, associated with a series of rather specific details. The designer then consults the files corresponding to the family he has chosen and looks for the specific design that comes closest to his order. Once he has found the similarity he is looking for, he takes the diagram of the existing die and adapts it to the new order. He re-dimensions a few measurements and then turns in a diagram of the new die to the programmer. When the order coincides with an extrusion belonging to a commonly used family, manufacturing methods have also already been defined and integrated into known methods. Manufacturing specifications combine the designer's diagram and the production plan devised by the programmer.

The concept of families serves to classify extrusions and to integrate feedback after adjustments. It includes the shape of extrusions, the shape of dies, and the results of the drawing process. It encompasses types of problems involved and types of solutions available. The specific history of each product has a bearing on the family it is classified in. Designers try to find out if the type of extrusion ordered has already been produced or what kinds of problems were encountered. They refer back to the families to identify similar cases. They check to see what problems were encountered and take note of them. They integrate the adjustments that were made on the tool in their calculations. They think in three dimensions. Designers gradually learn from this process. When considering an order, a designer begins not with the simplest design in an extrusion and die family but rather with the dominant design in the family. The design includes the history of results. Therefore, new calculations are unnecessary.

At Famiform, rules for designers are very precise. For example, they can take the form of equations developed from previous experience. A large number of dies with favorable results were measured and rules were established from them. These rules are used only if the design does not fit into one of the existing families. If it does fit into a family or resemble a known category, they use the old dimensions that can be found in the complex table of families and the archives of files and diagrams, which include modifications effected by the adjuster. At this point, the problems the designer is confronted with have a whole range of specific solutions. The more generalized rules are applicable only when the designer is working outside established families—that is, in less than 20 percent of cases.

A Culture of Calculation Rules

The design office at Reglocalc was founded only about 20 years ago, after a period of research during which basic equations and theories were developed. The design process is nothing like that at Famiform. At Reglocalc, the designs are created through calculation. The working hypothesis is that solutions can be arrived at that will thoroughly coincide with requirements. With this in view, the designers first began using computer applications to aid designers in dimensioning dies. It was their belief that the resulting calculations could be used on an industrial basis. After inputting a few basic parameters, the computer calculated the essential dimensions of the design.

Designers noticed that the calculations alone did not produce reliable designs. Adjusters on the press had to modify the die and effect major adjustments. In addition, the designers learned nothing from the experience, as the computer application did not allow for it. Each design had to start from scratch, that is to say, with only a few parameters to be fed into the computer application. Therefore, the results of former designs were not included in the calculations. No progress was being made. When adjusters said "This is not the way to go about it at all," designers could not listen to them; if they did, they could no longer use their computer applications. Finally, the designers and their managers concluded that it was best to get rid of the applications.

Instead, the Reglocalc designers adopted more flexible rules. They replaced the computer calculations with generalized design rules developed by the R&D department. As the rules are arrived at scientifically, designers can presumably rely on them. It is simply a question of applying them. They are generalized and free of specific problems. As opposed to Famiform, rules are used in all cases at Reglocalc. For each design, the designer starts from scratch, with the fundamental designs, parameters, and rules.

Apart from the head designer, who is responsible for reworking designs that come back from the adjuster, other designers are all considered on an equal par. No distinctions are made among senior designers, experienced designers, and other designers. Designs are processed in the order in which they arrive at the design office. When a designer finishes his work, he goes to the pile of new orders and takes the one at the top. In reality, this procedure is only theoretical. In practice, the designer quickly looks through the pile of orders and takes the order that best suits his capacities. Once he has obtained the specifications and the diagram of the extrusion, he asks himself how many openings and branches he must include in his design to achieve the correct shape. For example, in order to obtain a piece of square tubing, he includes four branches and four openings; for rectangular tubing, he includes three branches and three openings. This aspect of the design also depends on how many simultaneous flows there will be. Therefore, the designer defines the design of the die on the basis of a few generalized rules.

Then the designer's task consists of making a diagram while respecting and applying the design rules. Once the design is defined, he uses a diagram (saved on the computer) of a die having the same number of openings and branches as the one he wishes to design. However, this diagram is independent of the diagram of the extrusion for which the die was designed. It is also independent of the feedback received after adjustment (performance, adjustments made, etc.). The die design that is used at this point is therefore not linked in any way to its particular performance. It is presumed to be a viable design with respect to the extrusion ordered, regardless of the manufacturing environment of the drawing process.

Production feedback is processed through the head designer. He is familiar with adjustment problems and can therefore translate them in terms of necessary modifications to the diagram. At this point, the designer has only to modify the diagrams; he need not refer to production data. He applies generalized rules and works from existing diagrams; therefore, little creativity is called for. He is never asked to find a solution for doubling the drawing speed or reducing the number of adjustments needed on a die. If asked to design the same die, he simply reproduces his previous work.

Parallel to the computer files of die diagrams, the design office keeps complete files of extrusion diagrams, specifications, and die diagrams. However, to find the file that corresponds to a particular extrusion, the serial number of the extrusion must be known. These files are in chronological order. Therefore, if the designer knows neither the serial number of an extrusion nor the approximate date it was made, he has little chance of linking a die diagram to an extrusion diagram or to production feedback from the drawing process. The filing system therefore has an important impact on the way in which designers proceed. This filing system is itself conditioned by the culture of the design office: design is considered a science that can define the correct solution to a given problem by applying the rules of dimensioning. From this point of view, there is no reason to keep case histories on former designs or to refer to the extrusion initially ordered. In principle, therefore, a filing system is useless.

Reglocalc's culture of calculation and rules is ingrained in the objects they work with (the way in which complete files and die diagrams are stored on the computer) and in the memories of individuals working for the firm. The material translation of the culture of calculation and rules is all the more irreversible because designers are not aware of it. The design culture at Reglocalc becomes obvious when one attempts to introduce a new element.

To facilitate design work, the R&D department took it upon itself to study a great variety of situations and define guidelines or rules relating to orders of magnitude. These rules are used to make rough designs of new dies. From the point of view of the R&D department, for dies that closely resemble already existing ones it is better to refer back to the old die while integrating the feedback gleaned from experience. The rules are used only for completely new models.

However, in practice observation revealed that the rules established by the R&D department for rough designs are in fact adhered to very strictly. The designers adopt them as if they were actual law without ever questioning them. As long as they have applied the rules, they consider their design to be correct. Designers' faith in the validity of these rules is reinforced by their having been established by the R&D department (which is made up of scientifically oriented engineers) and by the fact that they coincide with the former conviction that designs should be based on calculations. Finally, the rules make it unnecessary for designers to refer to feedback data from production experience, which is difficult to do in any case because of the filing system. Their belief in the superiority of the R&D department prevents them from doing research and experimenting on their own. The culture of scientific calculation and rules is such a profoundly rooted tradition that it is nearly impossible to introduce objects linked to new approaches.

Starting from a computer that materialized undisputed principles of mathematics, the designers first completely relied on the computer that calculated and deduced a solution. When the computer applications were taken away (one day they were using computer applications for dimensioning and the next day they no longer had access to them), they reverted to their old system of logic and replaced the calculations with rules. Rules were considered perfect. There was no reason to verify that the applied rules worked in practice, or to question them at all. If the rules are applied, the results have to be good.

Switching from calculations to rules does, however, change the situation. Whereas the computer applications were rigid, rules have the advantage of allowing the design to evolve slowly as experience is acquired and difficulties are encountered. With the computer applications, the designers ended up with the same results if they input the same parameters. With the rules, if a design doesn't work, designers can attempt to go about it in a different way, and can modify the rules based on their experience. Therefore, though they do not question the rules, they cause them to evolve. The designers also evolve when, having failed to apply a rule, they happen to obtain a better result. They then become aware of the limited possibilities of the rules with regard to the problems that need solving. There is a true, but slow, evolution. The rule is not a trick of the trade, but a law. The rule must first be applied. Checking on the performance of former designs is not a reflex for these designers.

Defining Design Cultures

In this section we will note how different the design cultures of two design offices working in the same industry can be. These cultures are related to modes of thought as well as to objects (the table of families and its synoptic approach, computer applications, filing systems, etc.) and to practices (beginning the design process either by going first to the table of families or by applying generalized rules). Going still further, we will attempt to analyze how to make the implicit factors in the design process more tangible.

Defining the Rules

Basing the design process on rules, as is the case at Reglocalc, is very different from the design process at Famiform where designers attempt to grasp a general picture of how elements interact with regard to flow and extrusion. However, in both firms, designers rely on rules. At Famiform, only some of the rules have been formally written up, and these are only used under exceptional circumstances. At Reglocalc, the rules are constantly present and always used, but they are for the most part tacit and embodied in the process. When we are questioning designers, certain rules invariably come up. When we ask what their design process involves, however, their answers are always based on particular cases. They pick up whatever diagram is nearest them and relate the story of the design of this specific die. They do not give generalized answers. There are so many different elements to consider in each particular situation that it does not seem pertinent to them to speak in general terms in order to make themselves understood. Nevertheless, what they are able to relate about their particular cases is comprehensible only to those who are already familiar with it. They line up a string of characteristics and have great trouble in expressing themselves coherently. This is all the more true because they base their demonstration on a group of former characteristics and very few elements are clearly defined. When asked about these particular cases, why they do this or that, they begin to spout the rules. They attempt to justify their various design actions. This is when some of the rules they rely on become evident.

As we have seen, there are different types of rules. In the case of Reglocalc, certain rules have been in use for a long time, they date back to the time when designers relied on calculations. Other rules are proposed by the R&D department, and are often based on discussions engineers had with adjusters. Still other rules are based on production experience and feedback.

At times these rules conflict. For example, one rule stipulates that a die with few openings should have just as few branches; larger openings increase output but, on the other hand, reduce reliability. There are also other rules that stipulate that if openings have similar dimensions then flows will be more stable. Thus, the designer must depend on his own judgment to make compromises concerning the best solution to a given problem.

These implicit rules are often fairly coherent. At Reglocalc, designers think in two dimensions. They define a thickness for the branches, for example, and then in the diagram it appears to be a flat surface. Its depth or thickness is implicit. Yet adjusters know that a convex branch and a concave one will yield different results. Designers who think in two dimensions don't consider this aspect of the problem. Once the number of branches and their orientation has been defined, the overall structure is calculated on the computer. The computer calculates the size of all the openings and branches and their depth; the number of parameters to be taken into consideration is considerably reduced. The computer application implicitly calculates in two and a half dimensions—that is, in two dimensions plus depth. Once these parameters are obtained, the design of the die can be deduced. This is where the first set of rules, which is still in use, came from.

When Reglocalc did away with calculations, a second set of rules appeared. These are related to dimensioning. Once the number of branches has been decided, designers define sizes. Because they have designed so many dies and because they know whether the dies they made were stress-resistant, they are able to intuit the magnitude range of these dimensions. At least they can recall those that were not, and broke. Nevertheless, at the rate of one or two designs a day, this represents a good many designs to memorize by the end of a year. Designers starting out on their career ask their colleagues for advice. Later on, they develop their own tricks of the trade. Some of these tricks (for example, presuming that all branches have the same depth if the die is perfectly symmetrical) seem to be firmly based in logic. Other tricks of the trade (for example, presuming that width is linked to depth) are simply beliefs that may or may not reflect consensus among the designers. Sometimes tricks employ rules, such as "The effect of such and such a parameter is more consequential than the effect of some other." Still other tricks involve making compromises, based on experience, and then deducing other elements. Often, multiple compromises are possible. The precepts governing these tricks are not formalized. With the exception of a few generalized rules defined by the R&D department, the tricks of the trade stem from local beliefs or specific habits of clients. Thus, some drawers do not want to have openings that pass directly through the die. The designer therefore does not make a die with direct openings or the drawer will not buy the die, but he does not know why such openings should be avoided. Some rules are specific to one drawing factory. They simply become an institution among the chief adjusters. Thereby, designers proceed in different ways according to the drawers they work for. Still other rules are linked with production capacities.

At Reglocalc, the actors started from the principle that the computer calculated all dimensions. When computer applications were withdrawn, the designers replaced them with dimensioning rules without ever questioning the notion of two dimensions, plus depth. That is their culture.

At Famiform, though the designers deem the rules reliable, they still feel that they are less reliable than their customary practice of synoptic evaluation based on the vague notion of visual similarity. They don't believe that analytical evaluation, applying general rules one after the other, is pertinent; they feel that there are too many interdependent parameters. On the contrary, their reasoning is based on their intuitive grasp of a cohesive set of parameters. Though they have not always identified all the effects of their design process, they believe that concentrating their research within the framework of a specific family of problems and solutions will allow them to better evaluate former adjustments and ask pertinent questions. As their experience grows, the table of families and their corresponding files expands.

Experimenting with Intermediary Objects as an Indicator of Local Culture

Culture becomes more evident when new elements are introduced. At Reglocalc, for example, designers do not understand why the notion of families is important. If the notion is introduced, it either has no effect or is distorted. Nor do the designers understand why it would be useful to consult files, as they are convinced that their rules are reliable. This observation demonstrates that it is possible to indicate implicit modes of thought by introducing changes. We will therefore use this idea in our further analysis of implicit factors.

In intervening in the design process, we were able to make the designers' culture more tangible. This was accomplished mainly by analyzing elements of design methodology (based on our first exploratory hypotheses) and presenting them to the designers. They were asked to use the new methods in whatever way they wished. Soon afterward, they were consulted and asked what had become of the suggestions. Had they been used? In what way? Had they been diverted? If they had not been used, we wished to find out why. The working hypothesis was to presume that, if something was not used, it had not been explained correctly, it was not adapted to the problem, or the designer did not understand how it could be useful to him.

This procedure raised basic questions concerning design each time designers were consulted. Since the designers played their part well, if they left off the new approach, the hypothesis that something vital was at stake could be adopted. It was essential to discover what it was. A series of questions ensued that led to the discovery of a whole background of implicit factors, a culture, or a deeper understanding of things.

For example, regarding the design of difficult extrusions for which there is no reference family, the suggestion was made to start from a rough freehand sketch to find an idea that might work. However, there is a whole range of implicit factors in this suggestion and in the designers' reaction to it. Therefore, in suggesting that a sketch be made, the idea was to introduce a three-dimensional mode of thought, a new intermediary object between the extrusion diagram and the precise die diagram. It then became apparent that the designers deemed that there was absolutely no advantage to making a "sketch," especially by hand. The designers tried to use the intermediary object; however, from the beginning they thought of it in terms of two-dimensional drawings, and they found no logical reason to accomplish a rough sketch of this kind. This is how it became apparent that they implicitly thought in terms of two-dimensional diagrams, which led to the working hypothesis that, outside of the number of openings and branches and their depth, nothing else was considered, because nothing else fit into their two-dimensional mode of thought. The designers don't think in terms of three-dimensional coherence and overlapping effects of elements on one another. The experiment of introducing a new element brings out some of the implicit factors that influence the design process as well as our own hypotheses and the tacit convictions that led to our choice of new elements to be introduced.

Therefore, whenever innovative design methods are proposed, the manner in which designers imagine the objects they work with becomes apparent. This phenomenon cropped up again when attempting to introduce the notion of extrusion families. For this notion to be introduced, it was necessary to have a good deal of experience with various extrusions, which did not happen to be the case with the research engineer. Therefore, it was important that the designers themselves adopt the notion of families. However, whenever they began to do so, a new mode of implicit thinking came to the fore. For the research engineer, the word 'family' implicitly referred to the idea of a general three-dimensional similarity among extrusion types, problems, and solutions. The designers' response to the idea was to save the 2D diagrams on their computer and then refer to them in order to make diagrams of new dies more rapidly. Therefore, to the designers at Reglocalc, designing something signifies producing a diagram. To design quickly and efficiently, one simply reworks an existing diagram in applying the rules. It is not of primary importance for them to ask themselves if the die made from the diagram performed well in the drawing process.

The same observation was made with various intermediary objects that were created and tested. For example, the idea of making a skeleton diagram was introduced within the given magnitude range and dimensioned with the principal characteristics, as an intermediary step to designing the veritable object. The goal was to visualize, in a concrete fashion, the fundamental characteristics of the die and their effects. Then came the idea of architecture: defining the dimensions of the skeleton diagram without actually making a complete and detailed diagram. The third and last object would be the complete diagram of the die with all its characteristics. In the course of the experience, it became evident that these intermediary objects and their usefulness were not understood and were therefore quickly abandoned because design in this office is governed first and foremost by referring to generalized rules. For these designers, making a die is a gradual but direct and precise procedure. If necessary, they will modify one detail, but it is always something that has been pre-defined. There is no synoptic conception of the object based on intermediary objects.

Another example of an intermediary object: a formal document clearly defining adjustment feedback. This would mean storing diverse data such as the pressure used during the drawing process, drawing speed, and defects. Production feedback would also be stored: Was the die reliable? Did it perform well? How long did it take to break? The designers at Reglocalc do not understand what advantages such formalized feedback documentation would present because they do not refer to this type of data to fill a new order; they always refer to generalized rules for designing dies.

Similarly, an intermediary document could precisely define problems that need solving and identify the requirements that the die must fill (the price that the drawer is willing to pay for the die, the delivery date he will accept, the limited number of adjustments he agrees to effect, etc.). Designers tried to create such documents, but it represented an investment that did not show immediate profits. As they are under heavy pressure to be productive, such a major change in work habits is of secondary importance. Therefore, they never linked the problem to the solution, which is fundamental to the notion of families. In addition, radical changes such as this clash with routine procedure in designing product series. When designers attempt to innovate, it invariably leads to production problems because machine operators in the workshop are used to thinking in the context of the same culture, implicitly using two dimensions as a standard of reference. They do not think of the die manufacturing process in three-dimensional terms either.

The culture that emerged from testing design procedure hypotheses on objects is a legacy of past reliance on computer calculations that automatically generated programs for direct digital control. In this context, the personnel on the production line were not highly qualified; the computer managed everything, the programs were generated by research engineers and not by professionals in manufacturing that would have taken machine tool trajectories into account as well as cutting speed and a whole series of other factors relative to output and quality. Also, operators were rarely given the diagrams of the dies they were to make. They launched the program that the computer provided without always being able to verify that the program truly corresponded to the piece on order. Finally, neither the designers nor the people in the workshop had any control over the fact that the order of manufacturing operations affected the shape of the die being manufactured. We hypothesize that they failed to take this fact into consideration because they weren't aware of it.

Conclusion

The work of designers is ruled by implicit elements that interact in a particular way—that is, by a culture of design. These implicit elements stem from habits and modes of thought that individuals have assimilated. They are also reflected in the choices, the arrangements, and the use of diverse objects and tools (methods, technical standards, and design rules). They become particularly apparent when a change or a new element is introduced. Development and testing of new intermediary objects offers an opportunity to study processes of design that are unique to the office.

Operational Summary

1. Design work is ruled by implicit skills acquired through the long-term experience of designers and the collective experience of the firm for which they work. Above and beyond methods, technical standards, and scientific knowledge, each design office has distinct practices that depend on the nature of the products being designed, the specific needs of clients, and production constraints. They also differ in terms of other, much less tangible elements: implicit modes of thought and action rooted in tradition, a particular style, a culture, or a paradigm.

2. At least two design cultures can be identified. One is based on calculations and the application of rules; the other relies on a synoptic perception of shapes and problems.

3. Intermediary objects often reveal how problems are implicitly grasped and resolved. Contrary to conclusions drawn elsewhere (in chapter 3, for example), objects in this chapter are not active. Their significance and the manner in which they are used are essentially the results of implicit influences that must be deciphered by looking through and beyond objects.

4. In addition to analyzing the objects customarily used by designers, introducing new ones on an experimental basis reveals other tacit dimensions of design. When objects and tools conceived by the observer with a view to testing hypotheses concerning implicit factors in design practices (based on observation and previous conversations) are proposed to designers, the manner in which the designers react to these proposals (whether or not they use the suggested objects) is revealing.

5. Modes of thought and action are sometimes quite stable. As this chapter demonstrates, they even withstand the introduction of new objects, tools, and methods. This stability is all the more pronounced because designers are not aware of the specific nature of their mode of thought.

A Prototype Culture: Designing a Paint Atomizer
Éric Blanco

Chapter 5 explored the implicit aspects of design practices in two design offices. It revealed the existence of two design realms, one characterized by an analytic approach based on mathematical calculation and the application of general rules and the other by overall perception of shapes. In this chapter, we will present another design culture, in which prototypes play a leading role. Looking at a company called EPS, a subsidiary of an American group specializing in the development of paint spraying systems, we shall attempt to understand and explain the design practices that prevail in its design office. One of the main features of this department is the importance attached to the development and management of prototypes. We shall therefore pay special attention to the mediating role played by these objects in the socio-cognitive design process.

To get insight into the small world of this design office, we shall consider the design of a atomizer spindle to be used in painting installations in the automobile industry. This spindle sets a bellcup turning at a rate of 40,000 rpm, thereby generating centrifugal paint spraying. The paint is energized and, at 60,000 volts, is drawn to the grounded body of the vehicle to be painted.

In 1992, under pressure from customers in the automobile industry who wished to prolong the service life of atomizer spindles, EPS was obliged to replace the existing EPS1 atomizer with a new product. Moreover, EPS's competitors had recently launched an air-bearing spindle to replace a ball-bearing model. So the design process got underway, with those involved deciding in which technological direction they would go.

The Road from the Original Goal to the Final Development Is Full of Twists and Turns

First let us look at some of the main features of the design process. We will consider to what extent the design activities are dependent on a series of chance events that might prevent us from analyzing them as a linear, sequential process. The process is dotted with a series of prototypes, which we will meet along the way. They are just the tip of an iceberg which we will explore more closely.

Choosing a Technological Orientation

EPS decided to conduct studies with a view to developing an air-bearing spindle. A series of difficulties connected with this technological choice quickly became apparent. Bearings and thrust bearings may come into contact with other elements if the air supply is accidentally cut off.[1] In spite of the security mechanisms installed on the control panel, this possibility must be taken into account. The spindle must be capable of starting up again even after different parts have come into contact with one another at full speed. The traditionally used bronze bearings damage surfaces (friction, surfaces welded together due to the heat). As a result, when the spindle starts up again, guiding is less accurate, and the spindle may start to vibrate, which alters the quality of paint application.

EPS, in search of satisfactory technical solutions for air bearings, contacted Borg, a Scandinavian company that had developed an advanced technology. EPS conducted tests with the Borg spindle (air bearings made of porous graphite) but still had one doubt: graphite particles could break off and pollute the paint; on the other hand, if the paint polluted the bearing it would ruin it. The risk was too great. Together, EPS and Borg decided to investigate an alternative solution: a magnetic-bearing spindle.[2] Borg had, in fact, already developed a similar high-frequency-bearing spindle for the textile industry. A confidentiality agreement was signed, and the first prototype developed by Borg was handed over to the head of EPS's design office. It did not take long for EPS to take to the idea of the magnetic-bearing spindle. It would allow the company to make a number of significant improvements to the EPS1, in particular the feeding of paint from the center. Five years were then spent on development of this product in France.

Although the first prototypes served to validate the magnetic-bearing spindle, they were not suitable for paint application. For example, it should have been possible to remove the bellcup in order to clean it, but on the first prototype it was an integral part of the rotor. Moreover, the rubber legs on the spindle's shock absorption system did not withstand the aggressive atmosphere and the solvents used in painting facilities. EPS and Borg made a number of alterations. They were so substantial that, in the end, only the basic idea and the dimensions of the magnetic bearing remained intact.

While working on the Borg spindle, EPS was also developing an improved version of the EPS1 atomizer. EPS2 therefore came into being, with new materials and a significant reduction in the size of the paint, air, and solvent valves.

In addition, Borg was being restructured. Eiger, a company specializing in high-frequency pins for machine tools,[3] became the new owner of the patent filed by Borg. In 1993, a three-year contract was signed with EPS, stipulating that Eiger would supply these new magnetic-bearing spindles. The market launch took place in 1994. Review A of the specifications was drawn up in 1995 for the attention of Eiger. Nevertheless, the spindles were not yet totally reliable. There were still a few problems concerning the quality of the magnets. Deliveries were irregular, and the performance of some of the magnetic bearings was inadequate. EPS therefore contacted Labmag, a university laboratory specializing in magnetic systems. With Labmag's help, new and stronger magnets were fitted, increasing the bearing's stability. Nevertheless, the problems with the spindle poisoned the relationship between EPS and Eiger. After a series of faxes and meetings, EPS ended up redefining the technical specifications and setting up systematic inspections of all the spindles delivered.

Designing in a Turbulent Environment

In 1996, the three-year contract was coming to an end. EPS managers began to worry about the pressure they had put on Eiger, afraid that the latter would not renew the contract. EPS would then be unable to deliver the spindles to their customers. The management therefore created a "research" or study budget for developing another spindle based on variable magnetic reluctance,[4] as suggested by Labmag. Another contract was drawn up with this laboratory to define the measurements of this new system. The following development constraints were identified: the spindle had to be capable of replacing the Eiger spindle if the contract was not renewed; any connection with the Eiger patent was out of the question. Performance levels had to match those of the Eiger spindle.

Research therefore got underway, based on a large number of elements from the Eiger spindle. A first prototype of the variable-magnetic-reluctance spindle was ordered from EPS's usual precision tooling subcontractor, Usinalu. Test results were encouraging, but the industrial property consultancy Propindus found that the new spindle overlapped with the Eiger patent.

So EPS managers decided to approach the problem from a different angle. Taking the Eiger patent as a starting point, they looked for ways of getting away from it. Investigations into a single-magnet spindle were begun. Six different prototypes were developed. They worked, but inadequately. It didn't matter. They could be used in negotiations with Eiger.

In the meantime, the contract came to an end. Eiger didn't realize that it was automatically renewable. When Eiger's manager informed EPS that he did not wish to renew the contract, it was too late. Eiger found itself bound to EPS for a further 3 years. Now that EPS had the upper hand, they pressed Eiger to sell the patent. In spite of threats from the American branch of EPS to sue for astronomical compensation for the faulty spindles, Eiger did not move. As owners of the patent, they probably felt safe. Also, if EPS carried out their threats, they would no longer have any spindles to deliver to their customers. Now, as it happened, EPS pretended that they could do without Eiger at a moment's notice. To convince their supplier, EPS organized a demonstration of the single-magnet spindle on their premises. In addition, they gave Eiger the patent studies proving that this spindle did not interfere with their patent. EPS also showed the variable-magnetic-reluctance spindle, claiming "This one is even more efficient." The negotiations that ensued were difficult, but Eiger finally agreed to sell its patent at a quarter of the original price. Signing was scheduled for September 1997. Once the patent transfer had been agreed upon, the negotiating parties turned their attention to additional stipulations concerning, in particular, the settling of debts and new spindle deliveries.

In spite of this turnaround, EPS continued to develop the variable-magnetic-reluctance spindle. The results achieved with the first prototype were promising, and this type of spindle presented a number of advantages with respect to manufacturing. A pilot series was ordered from Usinalu, but, to everyone's surprise, the spindle's performance was nowhere near as good. In February, EPS decided to concentrate "definitively" on the permanent magnet spindle. The explanation for the pilot series' failure came in March, too late.

The Designer's Material Culture

Many objects mediating and representing the spindles of the future were passed around during the above-mentioned events. These objects included rough sketches, diagrams or plans developed using a CAD tool, prototypes, test parts, test and design review reports, specifications, faxes and in-house memos, invoices and order forms, patents, photos, and parts lists. These were the marks left by the process. All the documents pertaining to a given product have been stored in a "product file"; this "photo album" tells the product's story. It has, moreover, been re-structured, thanks to the introduction of quality assurance systems. However, this file alone does not cover the whole of a product's history: all the relevant documents are not painstakingly consigned to it, and there are other archives scattered around in various departments and companies. What is more, rough drafts and diagrams are not kept in this file. Dates and authors' names are often missing, which makes it even more difficult to reconstitute the design process.

In design offices like this one, it is not uncommon to find plans all over the place. Some desks disappear under a pile of plans, simply because the latter are fundamental to the design process. The widespread use of computer tools has not driven the plan to extinction: on paper, the designer can view the overall plan and its details at the same time. The computer only provides this possibility after much maneuvering, and in a series of consecutive screen images. It is therefore difficult to get rid of these graphic objects. After the introduction of quality assurance procedures, there should only be one approved and dated plan, serving as a reference. Design practices, coupled with the existence of photocopiers, constantly undermine this quality procedure. The proliferation of graphic objects is by no means specific to EPS's design office.

The disorderly accumulation of physical objects does, however, reflect one of the characteristic features of this design office and its design culture. Indeed, each designer has a multitude of parts, samples and prototypes hidden in his drawers: spindles, radial impellers, rotors, bellcups, and so on.

The Primacy of an Empirical Approach to Phenomena

The electrostatic spraying process is very complicated. The process has not been modeled, owing to the extreme complexity of the phenomena involved: high-voltage electrostatics, aeraulics at high rotation speeds, air flows in the painting booth, abrasion of the paint on bellcup surfaces, etc. Testing is the only way to assess theoretical solutions. In this respect, the designer's experience is to a large extent empirical, despite close cooperation with research teams. So far, the latter have only been able to confirm the phenomena observed by EPS's technicians, who would have liked them to indicate possible solutions. For example, the operators in one painting station realized that by coating the body of the atomizer with Vaseline they considerably reduced the number of paint stains on the atomizer. When the laboratory was questioned on the subject, it explained the phenomenon but was unable to suggest a simulation that could lead to new, practical solutions. The designers therefore came to this conclusion: "We have to try. We have to build a prototype."

A few years earlier, ideas took shape in the workshop first before they were tested; the designer developed the prototype and tested it himself. His expertise was based on this ability to create extremely short test-error loops between the drawing board and prototype testing. Nowadays, the development of prototypes is, to a great extent, contracted out. It has also become more complex. The design-development-test loop is longer. Nevertheless, it is still one of the outstanding features of this design office. The designer and the subcontractor in charge of developing the prototype are in almost continuous contact. Moreover, the designers are familiar with manufacturing processes.

The Specific Status of Prototypes

The prototypes, once they are back from the manufacturer's plant, do not follow standard part acceptance procedures. The "quality control" technicians who are supposed to check that they are compliant with plans merely set them aside for the design office. Besides, quality control is all the more difficult because the plans available to the controller are not up to date, as changes made during development are validated with the design office by telephone and are not added to the plans. Prototypes are modified many times during development. Therefore, thanks to time considerations, trust, and habit, the procedure defined by the company for the acceptance of orders is bypassed.

Technicians are, moreover, remarkably impatient when they are waiting for a prototype. As soon as it arrives, they will not rest until they have tested it to validate their theories. When a part arrives, they hunt down all the elements needed to put the prototype together: they take standard parts from the shop and tools from their co-workers; they go into offices and workshops to find what they want. They assemble the prototype in the laboratory and on the test bench. They check, first of all, that "it works." You can see the apprehension on their faces as they set the spindle up on the test bench. Seemingly relaxed glances and smiles betray their nervousness. Then they ask the laboratory technicians to carry out more advanced validation tests.

These tests are a fundamental part of the design process. Their purpose is to qualify and validate theoretical technological solutions. They also reveal problems that were not visible in the plans. Finally, they provide a glimpse of, and help to test, the manufacturing process.

Object Qualification Methods

In this design office, the qualification of ideas and theories depends, above all, on the qualification of objects. Designers, and the design office, spend a lot of time manipulating and testing them.

Obligatory Testing

Having developed a technical solution, the designer tries to measure its worth. He puts it to the test. The first of these tests, which takes place in EPS's design office, consists in answering the agonizing question: is the spindle stable? As far as the designers are concerned, this test is compulsory. Stability is necessary for applying paint correctly and avoiding damage to the spindle. If the prototype passes this test, it is considered suitable for service. It will then be required to pass a series of other tests.

To conduct this stability test, the technician sets the spindle up on a test bench and starts the rotor turning at, for example, 30,000 rpm. He then gives the atomizer body a sharp tap. He listens to the noise made by the spindle or lays his hand on the atomizer to see if it is vibrating. If the rotor has "disconnected" (i.e., started to vibrate), it means that the magnetic bearing's radial stiffness is not sufficient to hold the rotor in position in the event of a shock. The spindle is considered stable if it cannot be made to vibrate at 40,000 rpm.

When EPS's technicians were looking for alternatives to Eiger's magnetic technology, they had designed and developed a variable-magnetic-reluctance spindle. Stability tests on this spindle were mediocre, the stiffness of the magnetic bearing was judged inadequate. It was therefore decided that a new variable-magnetic-reluctance bearing would be developed, so that the two spindles could operate alternately. Simulations performed by the Labmag laboratory showed that, given the stiffness of the bearing, the "dumper"[5] system could be discarded; the bearing was therefore fitted, stiff, to the body of the spindle. Subsequent stability tests were positive.

The technicians therefore moved on to the "crash test." This involves cutting off the air supply to the bearing while the rotor is still turning. The bearing must be able to withstand this test several times without being damaged. This time, the results were bad. The glue used to plug the bearing's teeth melted and ceramic surfaces flaked. In spite of everything, it was decided that a pilot series would be launched to validate the mass production of the spindle. The pilot series was therefore developed, the glue being replaced by a more heat-resistant material. Unfortunately, the five prototypes tested were not very stable. The technicians could not understand why. The only way to elucidate the problem was to modify the spindles and perform further tests. As a result, the variable-magnetic-reluctance technology and the discarding of the "dumper" were dropped.

The Laboratory as an Organized Testing Mechanism
Tests play an extremely important role at EPS. Indeed, the "lab" occupies a predominant position in the spraying equipment design process. The designers from the design office spend a lot of time there. The following injunction is displayed on the laboratory door: "All non-laboratory users are requested to leave the equipment in good working order (including design office people)." This shows the extent to which members of the design office feel at home in the laboratory.

The laboratory can be divided into three areas. In the first area, all along the wall, are various experiments. Each experiment involves a fume hood with cascade water circulation and a paint catch tray. A control booth is used to drive a atomizer. Removable feed tanks are available for painting tests. "Test plates" pass in front of the atomizer on a conveyer. Two "powder" booths are located in the central area. The third area consists of a workshop equipped with the usual machine tools: bar-and-column drilling machines and lathes.

The laboratory is used in various situations. First, atomizer "demonstrations" are organized in the laboratory for visiting customers. Second, specific product tests are conducted there, for example when a customer wishes to test a new type of varnish in order to define optimal implementation conditions (flow, bellcup type, rotation speed, etc.). Finally, all the tests related to the design of new atomizers take place in the laboratory: endurance tests, adjustment tests, etc.

Just like the drawers in the design office, the laboratory's booths are home to a large number of prototypes and measuring instruments. Subsequent to ISO certification, the measuring instruments fall into two categories: those labeled ECME (for "control, measurement and test equipment") and those known as *indicators*. Each article is labeled with a number (if it is in the ECME category) or with the word 'indicator' (so that the article cannot be mixed up with an unlabeled ECME). The former are supposed to have been controlled and accepted, in terms of precision and their field of validity and application. The latter and not subjected to this kind of control. One of the consequences of this sort of classification is that, in a witty reaction to the laborious nature of some quality assurance procedures, some of the design office's graduated scales have been labeled as indicators.

The members of the laboratory are known as "escapees," as they are the people sent to the customer's facilities when painting problems arise.

The Personality of Technical Means

As it happens, my first job at EPS involved conducting tests with the different variable reluctance spindle prototypes available. The problem concerned the stability of the pilot series spindles. On grounds of availability, I went from one test bench to another to perform these tests. I was therefore able to note that each test bench had its own specific characteristics. I did not perform my first tests in the laboratory, but on the test bench in the production department before the spindles were sent out to the customers. This bench consists of a support surface attached to a sheet metal table. When someone taps the atomizer body to test its stability, the whole table shakes. I therefore observed that spindles are always more stable on this test bench than during other laboratory tests. Part of the energy created by the tap is dispelled by the table.

In the same way, laboratory experiments are subtly different. The experimental equipment on which stability tests are performed reacts in very different ways. All the more so because each technician has his own way of working: from sharply tapping the spindle to kicking the atomizer's support structure hard. A whole variety of spindle stability tests can be found in the laboratory. As a result, the conclusion that a spindle is stable is partially dependent on the men and equipment involved.

As a rule, experimental equipment is available to everyone, according to requirements. However, some equipment is less available than others because it is essentially taken over by members of the laboratory staff. Therefore, Laurent's equipment is really *his* equipment. Except in exceptional circumstances, he does not like it being used by anyone other than himself. Other equipment is reserved for customer visits, so I fell back on a supplementary machine used to train operators from companies where EPS spindles are used. This machine closely mirrors real operating conditions, except that it is not always in working order: People regularly take parts that they need from it. I therefore got into the habit of making sure that it was running properly the day before a new prototype arrived.

Prototypes and Bodily Commitment

Test results depend on the test bench and the person conducting the test. We have already seen the problems that arise from this in the stability test. This relativity also emerges in relation to data acquisition. The test is qualitative more than anything else. The only measurement and detection instruments are the operator's senses. He uses his hearing, for example, to detect unusual rotor vibration. Each spindle also has a specific sound: a variable reluctance spindle is not a permanent magnet spindle. With practice, a lot of things can be detected just by listening. Abnormal vibrations can also be detected through touching. The eyes are used to look for signs of contact or deformation. Overheating of the spindle's components is identified by the sense of smell.

As a result, it is difficult to pass test results, knowledge and know-how on. Describing a test result implies qualifying, in words, the sounds, tactile sensations and smells perceived. Consequently, it is understandable that the designers are so intent on performing the tests themselves. It is also understandable that, when a test report has to be drawn up, the parties concerned often gather around the test bench. They need to see, to hear, to touch, and to smell. I myself had difficulties in understanding until I did the tests for myself.

We quickly learn to detect anomalies through experience. It is less easy to classify spindle behavior on the basis of sensory perceptions. The problem is all the greater because test results are never clear-cut in the sense that the spindle does or does not vibrate. Some spindles vibrate at 20,000 rpm, others at 40,000. With some spindles, the vibrations die away immediately, others go on for a few seconds more and some disappear when the speed is reduced. Some spindles start vibrating when they are struck "sharply," others require less than that to vibrate. The so-called stability test reflects all these situations.

The Test as a Reflection of Operating Conditions

The role of the stability test is to make sure that the spindle is not likely to start vibrating during operation if it is subjected to a shock or unexpected stress. This test is warranted by the circumstances in which the product will be used. When I asked what this test was for, I was told that atomizers were sometimes subjected to shocks on site. What sort of shock, caused how and by whom? I never did find out. And yet, this test is judged necessary for all spindles.

I therefore turned my attention to the movements undergone by spindles when they are in use, especially accelerations, changes in the direction of machine arms, and rapid moves from one vehicle to another. These movements are liable to destabilize the spindle. I therefore conducted tests on the training machine involving substantial accelerations and changes in direction. Not one of the spindles tested started to vibrate. Now, spindles regularly come back from the customer's with friction marks, revealing that the axis has been put out of center. Unfortunately, we never get to know what happened, i.e. how the damage actually occurred. The customer never says anything; he hopes the guarantee will cover it.

The answer to the question "Is the spindle stable?" is difficult to find and always ambiguous. Yet it shapes the convictions on which a number of design choices will be based. It is used to decide whether another technical option will be adopted. A new spindle must be at least as efficient, and therefore stable, as the previous one. The stability test is the keystone to this assessment. It makes relatively high demands on the spindle—higher, it seems, than normal operating conditions would.

The Prototype, a Design Reference System

As we have seen, prototype testing is an important activity for designers. They rely on it to validate their design theories and to acquire a better understanding and knowledge of the object they are designing. The design activity, in this design office, draws largely on bodily resources. So what role do other resources play in the design process? For example, what is the relative importance of calculation and simulation in the designer's work? Prototype development and testing are based on a small number of preliminary calculations and simulations. However, we will see that the relationship between calculations and tests is always problematic. Nevertheless, relationships between designers and calculations are also complicated.

In the Beginning Was the Calculation
The designers from EPS are studying the Eiger spindle with a view to making a critical assessment of it. They suppose that Borg developed the spindle's geometry without prior calculation, by "guesswork." Wishing to shake off the bonding problems encountered during the assembly process, they plan to develop a one-piece rotor. They decide to perform a study of its measurements. They summon the subcontractor who usually handles anything concerning calculations and ask him to conduct the study.

To these designers, the calculation process operates like a black box. They trust their subcontractor, give him all the information relative to the problem, and expect to get results in exchange. The study report is, moreover, rather brief: 14 pages, four of which illustrate deformations of the rotor in different stress situations. It does not contain any mathematical theories. The geometry submitted to the subcontractor is, however, extremely limited by the fact that the different parts must be interchangeable with those of the previous spindle. There is, therefore, a whole series of parts which he is not allowed to touch. Moreover, a number of previous calculations revealed the impact of masses in the rotor, and the resulting deformations. The original shape cannot, therefore, be modified very much. The "calculation" subcontractor concludes his study with the statement that only two parameters have been selected. He therefore suggests different geometries and calculates measurements.

After a short meeting with EPS technicians, the geometry and measurements proposed by the calculation subcontractor are used to draw up the plans for the new spindle. It is clear that the designers have faith in the calculations. The study's arrival does not provoke the same sort of frenzy as the arrival of a prototype. Moreover, some calculations are never used. Unlike the prototype, the calculations are never put to the test. They are data, nothing more.

Then Came the Prototype

A designer is assigned to develop plans on the basis of the geometry proposed by the subcontractor. To do this, he uses the computerized calculation file. This however is not sufficient. He must also refer to the plans for existing spindles, which contain elements that he needs to reproduce in his drawing. But on which spindle should he base his work? Which plan? Being new in the design office, he finds it difficult finding his way around among the multitude of spindles, plans and plan versions. For manufacturing purposes he must, for example, define rotor specifications accurately, especially with regard to the Eiger spindle's magnetic unit. The difficulty lies in deciding which is the real Eiger spindle. He therefore starts to assemble information scattered about between the designers. He must also establish his own reference system, as each designer has his own version of the plans. He valiantly creates a base and an ad hoc reference system to do the job required of him.

As this is the first rotor that EPS has developed without Eiger's help, the entire flow-process grid must be drawn up. EPS wants, moreover, to master the whole process. This requires meetings between Usinalu and EPS. Design is not independent of manufacturing. Therefore, the drawing is based on discussions with machine manufacturers, as well as on calculations and the various versions of old spindle plans. Little by little, calculation becomes a remote, less decisive resource.

Although the plans have been drawn up, EPS's designers do not give up on the process. They go to the Usinalu factory to take a closer look at the manufacturing of the prototype, especially the bonding. Whereas they confidently delegated the calculation process, they make a point of watching each manufacturing operation—not because they do not trust the manufacturer, but because they do not want any of the details to escape them. Their attitude is clearly irregular: the manufacture of a calculation result and that of a prototype do not carry the same weight at all for them.

Two prototypes are built at Usinalu. During the process, the manufacturers modify the original plan in order to reduce the flaking of ceramics occurring during final rectifying or subsequent to knocks. Thus, design work is not confined to the design office and its designers; it is continued by the manufacturers. The resulting prototype crystallizes the adjustments made throughout the process. This means that when it is tested the whole process is tested, not just a single element or design theory. Both design decisions and manufacturing procedures are put to the test. Thus, if the prototype fails, it is difficult to ascribe responsibility to either the design process or the manufacturing process.

The prototypes have just arrived, and the designers want to try the new spindles right away. With their usual frenzy, they carry out the inevitable stability test. At 30,000 rpm, a strident noise stops them. They establish that the bearing has come into contact with the rotor. They dismantle the spindle and confirm, after a visual examination, that there has indeed been contact. This problem seems to be connected with the modification made by Usinalu. Yet this sort of problem did not come to light during calculations. The tests are inconsistent with the original calculations. The latter, even though they did predict that some elements might come into contact, did not anticipate it happening where it did. The designers conclude that something has escaped their attention—something that was not covered by the calculations either. They call the laboratory manager, debate, manipulate the prototype, and decide to cancel the modification made by the manufacturers. The design process continues not only during the manufacture of the prototype but also during testing.

To Get Back to the Calculations

Another test is performed and the problem apparently disappears. Yet no one can say why there was a problem or why it disappeared. The modeling of the rotor is, in fact, complicated. A number of aspects, such as how the glue will behave, are not clear and are therefore difficult to model. Besides, simulation results are very different to test results. Although the calculations showed that the prototype should work, the tests fail. The designers are at a loss. The tests are right, but it would be more reassuring if they understood why.

I work on further calculations with the members of a public research laboratory and present them to EPS. This triggers a defensive reaction from the calculation consultant, who is present. He is obviously worried that I am going to step on his toes. Once he has been reassured, discussions get underway, followed by comings and goings between the company and the public laboratory. Further calculations are made and fuel the debate. Together, we open the calculation black box. We end up exchanging calculation theories. As a result, EPS's designers get calculation results in general into perspective and acknowledge the necessity of tracking calculation processes more closely, especially when they are subcontracted.

The new calculations do not solve the problem. They have, nevertheless, simplified the problem and removed certain constraints. They have created new openings, thanks to an investigation into the sensitivity of certain parameters. These openings, however, require that a number of geometry-related constraints, deemed unchangeable, be dropped.

Calculation has become a new, unexplored territory. But the designers are pressed by time, and they decide to go back to their usual approach: build the prototype first and then test it. They put calculations aside, preferring to modify existing rough models at the smallest possible expense.

Operational Summary

1. Design is not a linear and sequential process.   There is an uncountable number of detours between the original goal and the final development. They are due to the exploratory aspect of the search for solutions, to the chance events that occur in an industrial context, and to updating of specifications after the discovery of new solutions.

2. Design results are affected by the incidents that occur during the design process and by the irreversible situations created along the way. The designer's actions and decisions are dependent on his company's strategy. He chooses technical solutions and prefers some performance evaluation criteria to others according to the constraints imposed on him. He takes this into account when investing in resources, thereby creating progressively irreversible situations. When new strategies are prepared and implemented, the new design is grafted onto developments already in progress; the designer does not start over each time the strategy changes. Consequently, the final design result is marked by the successive changes in strategy.

3. The ease with which graphic representations are reproduced is a threat to the harmonious representation of a given object.   Photocopying and multiplying computer files increases the number of rival representations of the same object.

4. The prototype occupies a central position in the design process when knowledge of the product can neither be formalized nor stabilized.   The development of a prototype therefore creates the possibility of testing. The prototype provides a possibility of testing the idea. Moreover, the idea sometimes emerges during the development of the prototype. Testing is a fundamental part of the design process: qualification and validation of solutions, building of knowledge of phenomena and products, understanding of the manufacturing process.

5. The development of physical models and prototypes fosters a close relationship and interactions between designers and manufacturers.   This relationship exists even when prototype manufacturing is subcontracted. In this case, the official prototype ordering and acceptance procedures are bypassed by designers and manufacturers.

6. The laboratory is an organized testing mechanism.   It uses the instruments and know-how assimilated by individuals. The personality of individuals and instruments is an important characteristic of this type of situation.

7. The testing period is a unique moment when products and their operating conditions, manufacturing procedures and manufacturers, test instruments, and laboratory staff are qualified.   Qualification may extend to the formalization and modeling of knowledge gained from testing.

8. The knowledge gained from the prototypes is also linked to the way the actors are physically involved with the objects.

9. The relationship between calculations and tests is always problematic.   Consequently, the relationship between designers and structural engineers is never easy. "Calculation," moreover, is as uncertain as testing. Formal calculation know-how and well-taught methods are not adequate when it comes to understanding the structural engineer's real work.

III

Technical Writing Practices

This part is devoted to a few specific practices, commonly observed in design offices and companies, that involve writing and graphical representation. Textual and graphical productions are particularly interesting intermediary objects. The practices surrounding their production, their circulation, their preservation, and their use teach us much about industrial design and innovation.

7

Writing Procedures: The Role of Quality Assurance Formats

Thomas Reverdy

In many companies, the writing of procedures involves management and engineering personnel. This practice relates especially to quality and environmental management. It has, moreover, become an essential component of industrial activity.

In this chapter, I attempt to systematize our understanding of technical writing processes. First I situate the question of formal writing. Using examples of procedure writing on two industrial sites as a starting point, I show how writing rules (formats, formalization) and the organization, standardization, and presentation of writing come into play. Finally, I consider the consequences of imposing an excessively rigid writing framework.

Formal Writing

The Attention Granted to Formal Rules

Industrial sociology is not used to studying the production of formal rules that provide an organizational framework and act as resources for company personnel. Industrial sociology endeavors to look beyond these rules, to explore how people interact, and to look at the informal rules that govern their behavior. The production of formal rules and management tools has been granted scant attention. However, other research conventions besides industrial sociology exist within the social sciences. They suggest that writing is an instrument of collective action, and that the modes and materiality of writing must be taken into account.[1] In this case, writing, circulation, archiving, and reading supports and practices must be studied closely.

With the establishment of quality control and environmental management, we are seeing a rehabilitation of formal writing within the company.

The new formal rules are known as "procedures." But is whether something new operating within the company, or is the change only superficial? Now, it is an established fact that the introduction of quality assurance standards has swept aside old rules and created a new literary order. The use of "standard formats" and document validation rules is increasingly widespread, and document distribution is recommended. Writing, and the organization and management of writing, enable us to "sort out" what already exists, and to clarify responsibilities. Writing instructions down, for example, provides an opportunity to settle disagreements about the correct way to handle equipment. As one operating technician said, writing sometimes produces a group reference: "When I first arrived, I copied out a co-worker's notebook: temperature, quantity, etc. Notebooks have practically disappeared. Now the guy looks at the procedures. Now, there's one, big notebook for the group." (Campinos and Marquette 1999)

In the same way, "records" (documents produced during periods of activity) provide a more reliable account of process performance than oral reports do. Finally, writing and the reorganization of information which goes with it enable us to adopt a more detached attitude and, therefore, to acquire new knowledge.

The Four Dimensions of Writing

The distribution of roles and the way in which writing is presented
Writing is associated with the distribution of roles between, for example, a requirements analyst and an engineer. Whereas the current context favors participatory management, some observers see in the introduction of quality assurance a return to the command-based relationship inherited from Taylorism. In fact, the distribution of roles, especially in writing, is fundamental, but the "analyst-engineer" model alone does not adequately explain it. Indeed, a number of writing mechanisms use a range of written documents and involve people with various roles. Members of a production staff often contribute to writing, which does not in itself guarantee the success of the operation. Contribution patterns vary, and the observer's role is to describe and assess them. Among the roles played is that of a "naive" individual who pretends not to understand or know anything yet asks essential questions. Besides contributing to the actual writing, various other people are asked to react to the written work in circulation.

The relationship with the standard: Inductive writing versus deductive writing
Writing has a complex relationship with the standards that define it. As far as quality assurance and environmental management are concerned, the writing of procedures is contingent on two requirements: on the one hand, it must comply with the standard; on the other hand, it must be a true reflection of the activity it describes. The first requirement cannot be circumvented if certification is to be obtained. The standard must be complied with. However, at the same time, quality assurance philosophy stresses the second requirement: that of "writing what we do, and doing what we write." This dual specification is clearly illustrated in the documents produced, some of which strictly comply with the standard (quality manuals, procedures) and some of which reflect activity (instructions). Both of these writing trends can be found in the field. One is deductive, a concrete expression of the standard; the other is inductive and attempts to abstract a broadly valid discourse from real activity. These trends are combined by the people involved. A problem arises if one of them is missing, for example: internal incoherence of documents, movement away from common practice, non-compliance with the standard. The combination of these two trends provides numerous writing possibilities, and this range of possibilities is reduced by the production and use of standard formats.

Formatting
Procedure writing is harrowing if there is no room to maneuvering—for example, if the standard is too succinct and does not shed enough light on what should be written, how it should be written, and in how much detail. Writers wonder what purpose their work is going to serve and if the auditor will accept it for certification. Therefore, guides and consultants are brought in to help implement the standard. However, the greatest consolation for writers is the possibility of recovering and copying tried and tested documents: general-purpose formats or model documents.

This practice of copying has not only produced a writing economy; it has also had unexpected consequences regarding document standardization in firms. There is, therefore, the risk that standard documents do not always express the diversity of local practices. So writers find it more difficult to appropriate their work, and, in trying to circumvent these difficulties, discredit the quality assurance system. Therefore the range of existing formats used, and the personality of the writers involved, have an impact on the nature, meaning and use of the documents produced. It is

therefore necessary to study the practices that attend formatting: enrollment of writers, appropriation and command of formats, organization of
writing time, circulation and approval of written work, update management, and so forth.

The relationship between written work and writing and action

People invest in writing when it answers industrial, technical, and managerial requirements. This relationship between writing and action merits closer attention. It is, however, complex. Sometimes the written
document plays a minor part in action, whereas the writing of this document was of major importance. Therefore, there is often a lot at stake during the writing process. For example, when a complex management
process has been misunderstood or when individual responsibilities are
poorly defined, setting the production process down on paper facilitates
comprehension and smoothes the way to an agreement regarding responsibilities. Other documents, for example work instructions, only come
into play once they have been written: their role is one of data memo-
rization, distribution, or compilation. The same is true for checklists,
forms, and guides. They help structure the exchange of information.

When written work does not meet expectations, it is often due to the
writing methods used—for example, when group activity is described in
a document written by one person (as often happened when it was
thought that a few skilled writers would do the job). The importance of
writing increases with the number of people involved in it: in the absence
of teamwork and without circulation of intermediary versions, writing is
in danger of being cut off from action.

Formats

Let us now examine the writing procedures at two industrial sites, paying
particular attention to the role played by formats. Quality assurance
serves as a vehicle for a small number of accessible, general-purpose formalisms, which proliferate inasmuch as writers draw their inspiration
from and copy them. However, they are highly simplified and, occasionally, unsuitable. They therefore produce labored writing.

When a company has set itself the goal of obtaining certification, it
tries to appropriate the standard. For example, a writer translates it, article by article, into a set of in-company procedures which are consistent
with other procedures and with the formal description of the company,
as well as with what are presumed to be the auditor's requirements. To

speed up the writing process, the writers use external resources such as quality manuals and ready-to-use procedures. They also enlist the help of consultants and draw inspiration from their writing methods. Alternatively, some companies focus on meeting their own technical and organizational requirements. They endeavor to describe, as best they can, methods which have proved to be the most satisfactory to them. They do not a priori work from pre-established formats. Yet in these companies, just as in the others, significant similarities emerge between different documents. This paradoxical observation indicates that standards, and occasionally company-specific standards, affect writing techniques beyond the requirements laid down by the standard.

Generic Rules

The way in which documents are organized corresponds to a set of rules which can be found throughout the industrial world, alongside quality assurance systems. Four types of document can be distinguished:

• the manual, a locally adapted translation of the articles that make up the standard

• procedures, which describe the main management processes across several departments and which generally comply with the articles in the standard

• instructions, recommendations, and operating manuals, which are specific to each department and which organize activity content and describe manufacturing processes

• records (e.g., records of analysis and measurement results, incident reports drawn up regularly or as a result of a specific event, inspection forms), which trace how things are done during and around an activity.

These documents are all structured in exactly the same way: aim, field of application, responsibilities, and so on.

Common Diagrams

Besides the general rules, various types of diagram are used. They make visualization easier, but also involve certain risks. Now, when a writer chooses a certain diagram, he is not always in possession of all the necessary information. Writers are not well acquainted with the different types of format, their possibilities, and their disadvantages. Besides, they rarely give any thought to the choice of the format before starting to write. In many cases, several formalisms are used in the same document or in the same diagram. This may make reading easier, or it may make it more difficult. The result, sometimes, is that documents are too detailed or incoherent.[2] The main formalisms used and their possibilities are as follows:

Types of Diagram

A *logical flowchart* (figure 1) illustrates linear processes made up of a series of decisions and acts. It is a basic quality assurance diagram used mainly to illustrate processes involving several departments.

A *tree* (figure 2) organizes events, objects, actions, or causes into a trunk and branches. This type of diagram is used when a group of elements have to be classified.

A *flow diagram* (figure 3) is made up of a series of nodes joined together by lines. The lines represent material flows. The beginning of each line corresponds to a piece of equipment from which a flow derives. Therefore, it is possible to visualize, for example, a heat exchanger's cooling circuits or a sewerage network. This type of diagram uses basic symbols and only illustrates homogeneous flows. It can be used to record flow rates

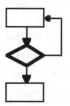

Figure 1

Figure 2

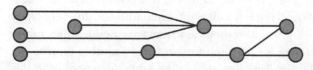

Figure 3

and calculate such things as dilution. It can also be used to detect sources of pollution, by following the flows back to their starting points.

A *simplified flow chart* (figure 4) illustrates main equipment, the control of which, in theory, has an effect on discharge. Each type of equipment is represented by a symbol. When linked together, they provide a simplified illustration of how the product is obtained. This type of diagram is mostly used to determine the position of measuring instruments and analysis points. Process details are not shown. These diagrams are created on relatively unstable, or ephemeral, supports: on a paper board, a page from an operating manual, or on the wall in the control room.

A *table* sets up systematic correspondence between two sets of headings. It is used to list analyses, instruments, process management recommendations, etc. It is relatively exhaustive, but it requires that the elements assembled be coherent. The writer must sometimes work minor page-setting miracles to produce heterogeneous situations within this format. The writer assumes that he can associate several lines and make use of the edges of spaces, but the reader does not always follow these directives. The result is that the table may be interpreted in varying ways, which are rarely clarified. In many cases, the reader pieces together the essence of a column by reading its content.

The Organization of Group Writing

Let us now turn to the organization of writing within a chemical company that is in the process of implementing the ISO 9002 standard.

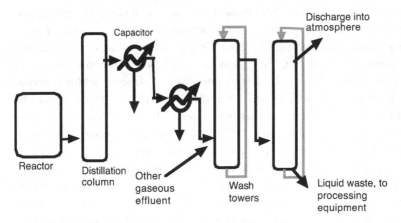

Figure 4

The Distribution of Writing and Document-Management Skills

The launching of the ISO 9002 certification project consisted of two days' training for managers and senior supervisors. The participants were amazed when management announced that no extra resources would be assigned to the project. "Quality" agents were appointed from among the supervisors. The documents were also drafted by supervisors, then approved by engineers. Today's good manager is supposed to feel comfortable with these new management techniques.

Nevertheless, the type of investment in the project differs from person to person. Those in charge of developing the quality assurance system discover and strengthen their convictions as the project progresses. Others see it as an opportunity to improve organization. Some people resist; others (even members of the engineering staff) acknowledge that they lack the requisite formalization skills. One engineer said: "When I look at an empty sheet, my mind goes blank. I don't know what to write." The implications and the understanding of these new management and writing techniques are affected by specific department rules and habits, by the distribution of power, by uncertainty, and by the old split between employer and employee. Some engineers and some supervisors refuse to put anything down in writing for fear of disclosing private information about their workshop: "We don't need your system, we handle things on our own." As managers, engineers discover that quality assurance rules are valuable as a management tool. So they have decided to extend their field of application, from the demonstration of product quality to "process control" and the management of facilities and administrative processes.

In this context, quality leaders make sure that they do not innovate too much and, rather, base their work on existing practices. Therefore, the implementation of a quality assurance system results in a new way of managing written documents. The operating manual replaces the multitude of operating documents previously used. These documents (e.g., personal notes jotted down by drivers and engineers) were seldom accessible; now they are collectivized. In addition, there are the notebooks in which daily instructions are written down. These documents are now managed by the quality system, which is accessible and up to date. Their purpose is to train operators and provide a reference in the event of abnormal operation. The operators are familiar with them, although few need them on a daily basis. They provide a structure for many activities. Formal writing management, therefore, affects many cross-company processes within companies where new writing, reading, and document-management skills are being developed.

Development of a New Interest in Discipline

The need for accuracy and discipline in writing results in a lot of adjustments among the people involved.

The degree of accuracy and discipline varies from document to document. Writers do not write everything. Each time they start to write, they choose a level of granularity beyond which they will not go. This level depends on the type of document, the writing requirements identified, and the extent to which the writer finds formalization worthwhile. The level of granularity increases in accord with the degree of conflict over individual responsibilities in the management process and in the event of accidental discharge. It is also important in certain inter-departmental "contractual" documents, such as the analysis control plan; tests which the workshops require the laboratory to perform must be listed accurately. Requirements that are meaningless to some (for example "performing such and such a test at such and such an interval" for the purpose of working out a specific calculation) are highly significant to others (because they involve a series of relatively complex actions or because they are used to check individual work). Thus if writing is not taken seriously, it may lead to conflict or result in preventive or reparative action with regard to other people. On the other hand, documents formalizing individual action which is already structured by a certain skill, experience, or trade are less precise.

The degree of discipline required from company personnel is often heightened by the anticipation of external auditor demands. Auditors and consultants sometimes encourage formalization beyond what appears to be necessary. When this happens, the process loses credibility.

The Documentary Format as an Instrument of Coordination

The need for discipline is twice as important when the documents play an essential role in coordinating and organizing work between departments. My first mission—updating the analysis control plan in the environmental department—brought me face to face with the reader's insistence on discipline in the writing process. The following story of a documentary format portrays the conditions in which it was updated and gives a fair idea of exactly how much has been invested in it.

Implementation of the ISO 9002 standard in certain workshops resulted in the establishment of a number of standard documents which management quickly recommended should be put into general use. The analysis control plan—a highly detailed table listing the analyses regularly needed by the workshop—is one of them. This table includes sampling conditions (who, where, how often), transport and analysis data (laboratory, analysis methods, chemist's qualifications), and distribution details (addressees, their location, and their telephone numbers). Previously, two types of document were used to manage analysis procedures: (1) the operating manual stating which samples were necessary and how they should be processed and (2) the procedures used to analyze samples in the laboratories. Between the two lay a zone of some good practices (for example, the chemist came by to take the sample) and a lot of disagreements (analyses which were missing or had been performed unnecessarily, undelivered samples, unfeasible analysis requests, results which were unavailable or which were overabundant). As soon as an engineer tried to obtain a new analysis or to get rid of an existing one, the whole system was upset. Therefore, it tended to remain unchanged, and the same errors occurred over and over again. To avoid this situation, the quality leader, along with a few operator and laboratory representatives, strove to create a single reference document, outlining responsibilities, deadlines, methods, a list of samples to be taken, the delivery and analysis of these samples, and the exploitation of results.

The operators and the laboratory gave this new format a positive welcome for two reasons: (1) there was a considerable need for clarification; (2) the chemist and the operator wrote the document together and signed it together. The time spent writing the document was a time of mutual discovery (respective restrictions and know-how) and explanation. Alongside the table, the operators added simplified diagrams of their process, locating the different sampling points. Thanks to this document, and to the process description (operating manual), operators are able to testify that they control the quality of the products manufactured.

In spite of this, the drawing up of each analytic control plan is still fastidious work and requires a lot of coming and going between the workshop, the environmental department, and the laboratory. It is both a contractual document between the laboratory and the operator and a detailed formalization of the work performed and the responsibilities involved. Its establishment is sometimes a prickly subject between the people involved. The laboratory, for example, has a keener eye for mistakes, as it has to carry out most of the work required. On the other hand, the operators accuse the laboratory of being persnickety and taking too long to correct the document.

To adjust the environmental department's analytic control plan, I took the table and went to see the people concerned several times. The writing process therefore proceeded in a series of mini-decisions, most of which were never clarified (verification of the necessity for each analysis, its frequency, and its feasibility). I thereby discovered the multiple uses to which this document was put, and the variety of things which were expected of it. I also understood that the format's success depended on how exhaustive it was, on its simplicity, on its high level of generality, and on the different ploys developed by the users to explain the diversity of situations. I also observed that the discipline employed in drawing up the documents was driven by a desire not only to be efficient, but also to respect the user.

I was also surprised by the extent of formalism surrounding this document. For example, the environmental engineer asked me to send a memorandum to the laboratory requesting that the values in the previous version be brought up to date. A few days later, I realized that nothing had been changed. In fact, quality control rules forbid this type of ad hoc updating and require that all copies of the new document be updated and signed by those who signed the previous version. Now, this kind of formalism is important for chemists, who need to have complete, up-to-date, reliable documents at their disposal. Those who criticize this bureaucracy are often individuals who do not have much interest in the documents, or writers who barely give any thought to how the documents are going to be used. Do not forget that laboratories are not given as much consideration as operating departments. The documentary format is a decisive resource for laboratory managers. Through it they are able to obtain a more regular definition of analysis needs from operators. It simplifies the internal management of their department.

Although the document is an instrument of coordination, it does not operate successfully on its own. The document-management system is an essential complement, as various cases of abnormal operation prove.[3] For example, the environmental technician noticed that certain analyses required for self-supervision purposes had not been performed. Now, the chemist who was supposed to perform these analyses did not have a copy of the right reference document. The environmental technician took it out on the chemist's manager in the following terms: "Can you tell me why your assistant hasn't carried out the analyses required of him?" In fact, the attack was aimed directly at the chemist's manager. He had an updated copy of the document and had not bothered to make a copy of it for the chemist who was supposed to use it on a daily basis. Control plans can be a new resource for the laboratory if this resource is not monopolized by the manager.

From Copying to Organized Standardization: The Distribution of Findings

Producing a work tool is a serious investment, even if it is only a straight-forward document. The aim is to produce something which will be widely used over a long period of time and in a variety of situations. Can such a tool, produced locally, be generically valid? There are plenty of examples of local, "makeshift" tools, or "findings," that subsequently acquire wide-spread importance. At the site studied, the environmental management apparatus was created through local experimentation and the adaptation of generic tools, then distributed to the whole site.

Abstracting a format from procedures drawn up by others is a common practice. It speeds up the process of creating new tools and writing doc-uments. This practice does, however, involve a certain risk: the "reprocessed" documents are often out of context; the writer does not know what their original purpose was, what their creators were aiming for, or what their period of validity is. Now, these documents contain implicit elements which can play nasty tricks on their users. The envi-ronmental department therefore attempted to organize document pro-duction along the following basic lines: problem analysis (formulation of the initial requirement), local experimentation (verification that the rule complies with existing "forms," local concessions regarding different requirements, etc.), distribution for widespread testing (among a few carefully selected engineers, with request for feedback), and adjustment and local translation of general formats. The aim was to try to harmonize processes through continually assessed action. One environmental man-ager said: "It's crazy how difficult it is to sell homogeneous factory processes. I'd rather avoid doing what was done for quality assurance, but people keep on complaining, contradicting, saying that its not applicable to them as it is."

The danger lies in forcing solutions on other people which are only valid locally. However, this danger is limited because the people involved are, in fact, cautious about the viability of generalization. Also, distribu-tion is progressive and is constantly adjusted, adapted, and fitted to local needs. Finally, a product based on an initial, local translation has more weight than a disembodied concept or an over-generalized format.

The success of written work depends on the scope of the adjustments made and on the common meaning that emerges from them as the doc-ument becomes stable. Now, iterative, group writing does involve a num-ber of irreversible elements. Its success, and the number of people who approve it, reflect a period of painful questioning. Besides being inter-subjective and conveying a common meaning, documents must be consistent with one another and with existing practices and tools. The document, now at the center of a network of human, technical, and textual resources, is increasingly independent of its creators' initial intentions. The action-based logic that prevailed when it was first created is therefore being obliterated by the demands of the writing itself and by the necessity of making the document consistent with other resources.

The Presentation of Writing

As the formalization of equipment management gets underway, two prominent tendencies emerge: (1) the inadequacy of existing knowledge supports and the difficulty in gathering them together and (2) the importance of how meetings are staged. By closely observing the writing process, it is possible to gauge both the importance of the socialization process and the need for supports to stimulate communication between the people involved.

How did the work groups approach the task of drawing up a list of equipment? In general, the meetings involved an engineer from the workshop, his assistants (foremen), the quality leader, the environmental engineer, and me (as a trainee.) The problem analysis that follows is interactive: what goes on in such meetings depends partly on how they are staged and partly on the people and the resources present.

Transitional Documents

In general, the discussion begins with a reminder of the agenda and of the questions to be addressed. Then, the project manager mentions one or two cases of polluting discharge. Often he bases his presentation on one document or another and the other participants request a copy of it. The following items may be found among these documents:

- a list of discharge points (drawn up for the occasion), classed according to the nature or the structure of the pollutants

- an illustration and, possibly, a diagrammatic table of the workshop's layout, an outline of the Analytic Control Plan (ACP)

- a diagram showing discharge into the atmosphere or the water system (drawn up by a trainee to provide an exhaustive list of discharge points and assess their compliance with the decree of March 1, 1993)

- a process outline, which may be detailed (a list of equipment) or not (main workshop flows). This outline was drawn up in a different context and for another purpose, such as the improvement of a piece of equipment or the preparation of a new investment.

• a list of so-called vital equipment. A classification method based on a variety of criteria was used to draw up two lists, one of "vital" equipment and of "important" equipment. Their main purpose is to determine upkeep and maintenance modes. If a piece of "vital" equipment breaks down, it is supposed to be repaired immediately; the critical parts are in stock.

The choice of this document is seldom insignificant: it provides a satisfactory picture of how knowledge stands at present and is easily available in the workshop. A single document is sufficient to begin discussions. A large amount of time is given over to interpreting and updating the document. The participants do not modify its "scale." Everyone tries to note the modifications brought up during the discussion on his copy of the document. The participants frequently bring up the fact that summary documents, which consider the issue of discharge control from all angles, are seldom available. The document presented often reflects the progress of the project. In many cases, the meeting is more an opportunity to start the project off on a group level (involvement of supervisors) than to assess the document which is up for approval.

It is surprising, nevertheless, that no figures on discharge are provided at these meetings. They are calculated in the environmental department and then distributed to the engineers concerned. Now, nobody brings them to the meeting, as if a rough idea of them were enough. I did not even know they existed.

This drawing up of an equipment list relating to the management of polluted discharge is based on documents drafted in accordance with a variety of decisions made in different contexts. These documents make the transition between existing knowledge and new knowledge possible. New knowledge is, to a certain extent, predetermined by the type of document used to open the proceedings (including its content and its deficiencies). From then on, group activity takes on a momentum as the protagonists strive to produce a concise demonstration of their discharge management techniques, based on scientific reasoning and the improvement of knowledge. The demonstration must be "scientific" and well argued and the documents are in fact used to back it up.

The Skills of a Good Naive Participant
The initial objective of these meetings is to make sure that the list of equipment is complete. There is a tacit understanding that the list of discharge

points will also be verified and that decisions regarding analyses and management tools will be approved. The naive participants ask questions and challenge accepted ideas. The operators present their work, their decisions, their arguments, their information, and also their questions.

There were two of us playing the role of naive protagonist, one fake and one genuine. As I am truly naive, I can only ask relatively standard questions, consistent with the decisions taken during previous project management meetings. However, my questions may come to a sudden end if my interlocutors do not take up the questioning themselves, if I myself do not have enough "weight" or enough knowledge which can be turned into questions, or if the other "naive" protagonist, the quality leader, does not revive the investigation with new questions. The quality leader is not truly naive—he pretends not to know; infinitely cautious, he stands back and lets the others (the unit manager and his assistants, the environmental engineer) show that they know what they are talking about. However, he has worked on the site for almost 40 years, including 10 years in the quality department, so he is familiar with each workshop, its operating problems, its organization, and the behavior of its members and managers.

Making use of such influence is not easy for me. The only information I have at my disposal concerns recent events which, due to their gravity, received a certain amount of publicity. So the investigation relies on the participants' memory, which is relatively disparate according to their personal experience and the extent to which events were publicized. Whenever a difficulty required explanation, the protagonists tended to use events everyone knew about. Therefore, controversies arose over situations which had been avoided during meetings as the people present were not acquainted with the subject and were not able to revive questioning.

This biased judgment of a systematic and formal description, spurred on by the naive questions of certain members of the department, seems unusual to a lot of people: indeed, it challenges a certain natural opaqueness surrounding the management and daily running of their workshop. Most of the time, these practices are only brought up to date if they have a noticeable impact, such as a variation in the concentration of pollutants.

An engineer is reluctant to address various delicate questions. He asks: "Are the auditors going to get a close look at the workshop, the equipment . . . ? If they are, and if they're well up on it, they're going to find loads of things that I have not written down." He then provides a list of small controlled experiments and minor short-term discharges. The engineer reveals that he has a precise knowledge of this discharge, which the environmental department was unaware of because it was partially hidden by the effects of dilution and compensated for by discharge from other equipment. After discussing these various emissions, he concludes: "We must bring everything out into the open, show that our work is exhaustive, even if we do have a few arrangements for salvaging things. We have to explain, specify what we do."

Apart from one engineer who fiercely defended his managerial independence, all the participants worked together, playing the game. Engineers and supervisors went along with it and provided a large amount of information, encouraged by "respectful" questioning. Thus, group writing depends on how it is organized and presented.

The On-Stage Drama

The meetings were, on the whole, successful in that the most problematic subjects were addressed. They failed once, producing a dangerous misunderstanding. This failure can be explained by the ways in which different methods of thinking and acting interact, the relationship to the various documents and aids, and the configuration of the game during the meeting.

During one meeting, the participants insisted on the importance of the device used to analyze the carbon monoxide content in the steam from the natural gas boiler. In their opinion, this piece of equipment should appear in the Plan de Contrôle Instrumental (PCI) as an environmental control tool. The quality leader believed that every important piece of equipment should be systematically registered in the PCI. The environmental manager argued that this equipment played an important part in controlling energy consumption, which is one of the environmental department's concerns. The supervisor thought, on the contrary, that this equipment came under daily operation and the optimization of the oxygen supply. He refused to see it registered in the PCI, which would have implied a formalization of its calibration.

The head engineer (who was absent and was replaced at the meeting by his supervisor) reacted very quickly when he saw the minutes. A discussion between him and the environmental engineer ensued:

I'm not at all satisfied with your report, you pay too much attention to completely insignificant aspects and don't mention the most important. What I've heard from the supervisor isn't very good either.

But we followed what the supervisor showed us.

The supervisor didn't show you anything. He answered your questions. He felt as if he were under attack, being put to the test. In the meantime, you paid too much attention to totally meaningless discharge. Why attach so much importance to it? Carbon monoxide comes under process control, my boiler's yield depends on it. . . .

In fact, during the meeting, the supervisor stood up and went to the board to explain how the boiler works. He left his documents behind. As he stood in front of the board, without any documents and not very sure of himself, he let himself be guided by the other participants. The information that he provided sparked off a series of questions which were beyond him. On top of this, the rules of the game had not been fixed. Nothing had been defined—not the environment (Is energy consumption an environmental concern?), and not what the PCI was expected to contain (Should all important equipment be listed, or only the equipment necessary to quality and environmental control?).

Excessively Structured Writing Becomes "Deductive" Writing

I had the opportunity to observe the implementation of the ISO 14001 standard (environmental management) in a second company, working in the mechanical engineering sector. The members of the quality team saw the project as a chance to experiment with new writing skills within a limited framework before extending them to the rest of the quality department. Following this line of argument, a two-day training course on procedure writing was organized by AFNOR (the French standardization organization) with the aim of "launching the ISO 14001 project." The description of this training course gives insight into a different way of managing writing activities. Some of the aspects that emerge have already been analyzed, and they differ greatly in impact.

Criticism of Excessive Formalization
Criticism of the company formalization practices established by the quality department focuses on the following: excessively long procedures, superfluous text with the diagrams, difficulties in detecting the management processes, over-complicated updates, etc. This explains why consultants and AFNOR insist, at present, on a trimming down of documentation and take responsibility for any excess formalism. However, this new way of seeing things underestimates how difficult it is for writers to acquire formalization skills. Formalizing through diagrams is no easier than writing.

Organizing Writing

Not only is this method of formalization complicated; it also removes content from procedures. Therefore, the program for the two days at AFNOR includes rapid training in how to create logic diagrams and the drawing up of a procedure for each article of the ISO 14001 standard in the shape of a logic diagram. The participants are motivated young engineers and technicians who will be involved in the ISO 14001 project and in the rewriting of the quality system. Most of them are novices in environmental matters. Groups are created for the drawing up of each procedure. The process promises to be extremely participatory.

The proposed writing method requires simplicity and conciseness. After identifying the main environmental management processes, it suggests formalizing them in the form of a logic diagram: splitting the different processes up, then defining the input and output data for each one. Then, the different actions and decisions involved in the process, from input to output, including all the necessary loops, must be described in detail. A leader and a group of participants must be appointed for each action; a decision maker and a number of people to be consulted or informed must be named for each decision. The required documents, instructions, and recordings must be cited. Except for this logic diagram, the procedure must be succinct.

During the first day, each group is put in charge of one of the articles of the standard: environmental aspects, legal and other requirements, in-house communication, supervision, and measuring. The writers work individually. This, however, gives rise to the first problem: splitting the standard into chapters in order to draw up a set of procedures presupposes that it is possible to define a distinct management process for each article. Each group adheres scrupulously to the method. The first step is to read the article and identify the inputs and outputs of the process described in it. For the first article, the input consists of a list of environmental aspects; the output, a list of significant aspects and effects. The group gives a detailed account of the process from input to output, in compliance with the grammatical structures used. One of the participants then explains the logic diagram: "First of all, the list of environmental aspects is brought up to date, taking into account statutory and process modifications. Then a second list, of controllable aspects, is drawn up. A third list enumerates significant aspects and a fourth, significant effects." Once the process has been defined in detail, the distribution of responsibilities is discussed. The decision as to what is or is not controllable falls to the site manager: "Since he is the most familiar with production requirements, isn't he the most likely to know what is feasible?"

The second group uses the same rules to draw up its procedure (legal and other requirements). It deliberates on the inputs and outputs. The consultant, who is moving around between the groups, suggests that "the process output consists of applicable requirements." The group then decides that the process input comprises statutory changes or the list of environmental aspects. The group transforms the issue of access to requirements into a simple matter of the circulation of information within the company. The distribution of responsibilities is also discussed: the central environmental manager is in charge of legal aspects; the site environmental manager is responsible for rating conformity and instigating corrective measures. Indeed, because the company is located in a number of different places, the distribution of responsibilities between the central environmental department and the factory environmental departments is an important feature of the project.

Meaningless Procedures

Written this way, the two procedures are superfluous and contradictory. Still more alarming, they do not explain how controllable or significant elements are defined. Neither do they tell us where requirements come from, or how regulations are taken into account. Every time the groups try to include significant information, the consultant argues: "You've got to keep it on a company level. We don't care about sources of information, we'll go into that in another document. You are formalizing know-how, and not the company. It's perfectly natural, everyone makes the same mistake." In other words, there is no need to know, at this level, what an environmental aspect is. It will be defined in another document. On the other hand, three different lists of environmental aspects must be drawn up, according to whether the aspects are controllable or significant. The process definition must be based solely on the grammar used in the standard.

A discussion about the distribution of responsibilities gets underway, but it is based on a process that lacks real meaning or content. The participants do not know what skills are required for each step. The main achievement is an insight into the exchange of information between central and site environmental managers. However, no details are revealed regarding the nature of this exchange.

At the end of the day, when events have been summed up, a certain degree of frustration can be sensed. To be sure, the participants have tested the new writing method and have understood it. They have also grasped the fact that the standard is complex and superfluous. This becomes even more apparent when one tries to define a logic diagram for each article. But they feel that they have not addressed the real questions raised by the standard, and they are not satisfied with the documents written.

The environmental manager, on the other hand, had drawn up a rough outline of an environmental analysis procedure, including a list of environmental aspects. It would have been interesting to use this work; it would have given meaning to the standard's abstract designations. The environmental manager did not think it was worthwhile using this outline as a basis for discussion. He thought it best to take the standard as a starting point. His work was innovative, but he was not very sure of his choices. He preferred not to impose a document that, besides being intermediary, would be a debatable basis for future work. Moreover, some guidelines on the application of management standards recommend working from existing processes and documents, and formalizing them before checking if they comply with the standard. Now, in this company, management documents and systems are plentiful. It would have been preferable to work together on the range of existing documents (water, environmental, and waste analyses) to define a process compliant with the standard.

The lack of skilled environmental representation, the sharing of writing activities between subgroups, and the difficulties involved in learning new formalisms result in the development of abstract formalization practices. None of the company's skills are clearly illustrated in the logic diagrams. The system neutralizes the participants and imposes the standard's abstract requirements.

Operational Summary

1. The requirement to write "what we do" is so vague that writers need other guidelines, implicit or explicit. These guidelines provide a framework for their writing.

2. These guidelines are rarely chosen with full knowledge of all the facts. These choices not only reveal writing mechanisms; they also provide a structure for these mechanisms.

3. Copying and harmonization of existing formats are essential. We can therefore understand the importance that writers attach to these documents and written records, and appreciate why they are willing to invest (formally) in them.

4. Frameworks differ.   A good understanding of these frameworks and of writing routines is a skill that should be brought forward and used in the development of quality assurance and environmental management systems. This skill allows writers to appropriate the standard and to use it as a separate learning tool (Segrestin 1997).

8

The Role of Graphical Representations in Inter-Professional Cooperation

Pascal Laureillard and Dominique Vinck

This chapter concerns the role of the many graphical representations that punctuate a machine part design project. It pays special attention to the various forms of inter-professional cooperation (design, forging, machining) that develop around these graphic objects. It shows that graphical representations are closely connected with design organization, and that they are both the product and the ingredients of this activity. These graphic objects reflect the intentions, methods, and constraints of those who designed them. Therefore, they provide a good introduction to the project and its progress. Finally, the chapter shows that inter-professional cooperation is built on, among other things, the redesign of a certain number of these graphic objects. It is suggested that some of them be called "cooperative agents."

Graphical Representations in Action

Scene 1: A Substitute for Official Commitment

After a hiatus of more than a year, design work on the X01 axle is getting underway again within the company. The project managers had previously stopped the work, with the view of implementing a "do or buy" approach. In other words, several suppliers were consulted and brought into competition with the company's own design and manufacturing procedures.

Two schools of thought exist within the company. One holds that the future lies in the development of a subcontracting policy, through the purchase of the majority of vehicle components. This implies that the company would drop its in-house manufacturing activities and become an assembler. The second school of thought encourages development or preservation of in-house industrial operations. These two opposing trends throw project development into some confusion. They are also sources of social tension: there were demonstrations and strikes when it was announced that the engine for the new line of vehicles would be subcontracted.

The hunt for a potential X01 axle supplier was difficult. Several world-wide market studies failed to turn up a single supplier capable of meeting the company's cost and quality objectives. At the end of the day, the in-house procedure seems to be the most effective. Nevertheless, this conclusion does not take hold easily. The project managers' official decision to resume in-company design work comes very late in the day.

The members of the design department unofficially resume work in an atmosphere of confusion, without the written authorization of the project manager, but aware that they have to get started right away. Moreover, several of their immediate superiors verbally request that design begin again. Procedures do not, therefore, become official because management gives written authorization. The existence of two rough, solid models[1] of the mechanism makes them official. The fact that these graphical representations exist and have been discussed is sufficient testimony to the firm's de facto commitment to the project.

Scene 2: A Vehicle for Command-Based Relationships and a Partner in Innovation

The graphical representations produced and distributed will be discussed, analyzed, and explained once again by the people involved. Through these models, they discover not only design choices and the implicit theories of the other people involved, but also their new position within the organization.

Therefore, new technical data are introduced when the project starts up again. Two suppliers are approached by the management, and each puts forward a solid model of the brake caliper. This exasperates the designers. This competition from outside suppliers, plus the fact that the final choice will not be made until the development process is underway, according to who provides the best value for money, increases their work load. They will have to work on two studies at the same time until a supplier has been chosen. From their point of view, this situation also discredits the project managers.

The two suppliers deliver their solid models. A designer from the design department adjusts them so that they comply with his own graphic objects (old, two-dimensional plans). This operation brings to light a number of distinctive features in the suppliers' design. The calipers' unusual attachment system provokes discussions in the design depart-

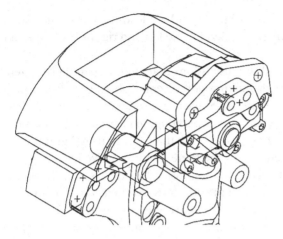

Figure 1
Solid model of the caliper.

ment. Designers wonder, in particular, how the caliper will behave when in operation. The design department had created an attachment system that allowed the caliper to follow the brake disk even when the latter was bent out of shape. Apparently, the suppliers work on the opposite theory: develop very rigid attachments which cannot lose their shape. This situation illustrates that graphic objects convey and express their authors' theories and intentions. The latter, however, are only revealed when a new graphic object is adjusted to fit a better-known graphic reference, and when the solution is compared with an alternative.

This situation also reveals, to the people involved, a change in the command structure. The members of the design department, who are used to imposing their design choices on suppliers whom they see as mere service providers, are seeing signs of a disturbing development. For the first time, suppliers are imposing their design choices on the design department. When the solid models presented by the outside suppliers are analyzed, the design department finds that the company's own hierarchy looks on it as a service provider. It now has to submit to orders from the suppliers.

Examination and analysis of the graphical representations prompt new ideas among the designers. Taking the caliper models provided by the suppliers as a starting point, they reach the conclusion that the spindle[2] to which the caliper will be attached must be redesigned. They suggest a new shape, illustrated by a few simplified drawings which give rise to further discussions regarding the optimality of this solution and the feasibility of such a shape in the forge. Looking at the drawing, they also note the symmetry of the solution (figure 2), which leads them to believe that only one type of part will be needed for the left and right wheels, instead of two.

Graphic operations and the discussions that attend them thereby generate a design and analysis melting pot, a center of adjustment for people and technical elements. Graphical representations display new "holds,"[3] allowing new constraints (operating, manufacturing) and intuitions to be examined. These holds bring the constraints to light; the latter are not listed beforehand. In the same way, innovative ideas (such as using one type of part for all wheels, instead of two) spring from the graphic solution and not from a previously established company strategy (to reduce, in general, the amount of part reference numbers).

The drawing also provokes reactions. As soon as a person sees it, he latches onto the holds, which he then subjects to his own point of view, represented by the spectator. The assembly planner[4] pays a visit to the design department. When he sees the drawing of the spindle for the first time, he wonders whether it will be possible to assemble such an architecture. He usually fits the kingpin after the brake caliper. It will be impossible to do so in this case. As the caliper is situated above the kingpin, the axle cannot be fitted with the tools used for the existing range of axles. The possibility and the relevance of developing a new assembly plan, which could lead to alternative solutions (preventing overhang phenomena), is considered. The drawing, seen by chance, raises questions from an observer who is skilled in a different area. These questions lead the people involved to picture other solutions, and their consequences. Gradually, other questions are raised. Occasionally, a new idea concerning

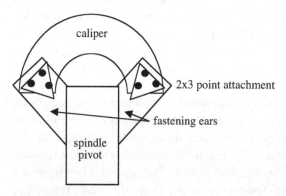

Figure 2
Side view of the new spindle.

a problem completely different from the one that triggered this chain reaction comes to light.

The chain reaction is not systematic. When two smiths pay a chance visit to the design department, they take a look at the first sketch of the spindle. When they see how the spindle's two eye mountings are designed, one of them raises the question of how the metal slug will be positioned in the forging die, to ensure that filling is correct and that the metal fibers lie in the right direction. He asks the question, but does not provide an immediate answer. The process comes to a temporary halt when the first question about the drawing is asked.

Scene 3: The Resistance of Reality and the Co-Production of Knowledge

The manipulation of drawings really puts design work to the test. A designer from the design department starts to polish up the rough sketch of the spindle on a two-dimensional plane. He tries to make the assembly more compact than the old design by inserting the brake caliper a bit further into the wheel rim (5 mm translation). He thereby gains 9.5 mm at the brake cylinder. He then notices that there is interference between the caliper and the rim size determined by Michelin. He also observes that the caliper's attachment surfaces are placed in front of the bearing seat surface, which is uncommon in the architecture of existing spindles (figure 3). Creating the drawing produces at least as much knowledge as the thought process that precedes it and the reading that follows it.

Graphic production is intrinsically manual and cognitive. The designer places an ABS (anti-lock braking system) sensor on the spindle. He thinks, chooses, and sometimes explains what he is doing while he is doing it. He takes account, for example, of the ABS supplier's recommendations. He chooses what he thinks is the most favorable position in relation to foreseeable problems (e.g., avoiding contact upon maximum wheel lock or moving away from the source of heat generated by the caliper upon braking). He pursues the idea of moving the steering lock hole away from the ABS hole in order to make the spindle shaft rigid. He also takes various constraints into account (for example, the positioning of the ABS cable). He constantly draws on his knowledge—for example, the shaft must be correctly fitted into the spindle; if a hole is made on either side of the shell, the shaft will not be rigid enough; this rigidity must be limited by the thickness of the shell, as the latter may create a significant number of constraints at the blending radius between the shaft and the plate.

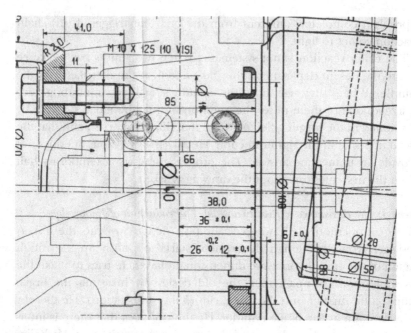

Figure 3
2-D overall plan.

Graphically Aided Cooperation

Two months later, the designers from the design department ask the machining[5] method planner to come and see them to sort out a number of design issues. They ask one another questions while the graphical representations are being passed around and examined. The designers ask, for example, if the fact that the caliper's eye mountings are curved on the shaft side is inconvenient. The method planner asks the design department to confirm that the B-spline curve that connects the bearing seat to the spindle shaft is identical to that used in previous studies. He also raises questions about the shape of the enclosure (figure 4).

Some questions are answered immediately. Others are deferred, in which case they point out the work that must be done and the information that must be provided (for example, measure the machine tool to be used, perform investigations or simulations) before an answer can be given. Together, the designers and the method planner also discover a variety of problems, as well as new possibilities (regarding, for example, the assembly of the part). Occasionally, they make a hasty sketch. Thus,

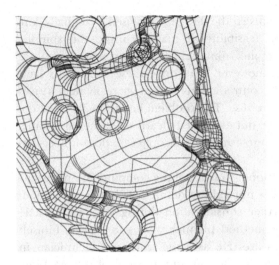

Figure 4
The spindle enclosure.

the people involved act on a number of local arrangements, sometimes as a result of a chance meeting, sometimes on the basis of a request made by a member of the group.

These meetings between professions are often spontaneous and informal. However, as the project progresses, more and more questions that require group decisions are asked.[6] The problem then arises of how to reconcile different viewpoints and heterogeneous information. Distribution of solid models and the accompanying observation sheets does not suffice. The people concerned need to meet informally, preferably in front of a computer screen.

Scene 4: Cooperating in Front of a Workstation or a Machine

The machining method planner comes to the design department, with a technician, to work with the interface agent on a computer. The interface agent tries to create his own computerized representation of the method planner's reasoning. In fact, the planner is not prepared for this meeting; he has not looked at the plans again since he established them with the sector manager. He is currently working on three projects, including the acceptance of machines and the input of digital data for the development of the axle-dedicated machine. He has little time for working on the X01 project. Present in the workshop and available for the design department at the same time, he must have a talent for ubiquity.

The subjects broached concern the difficulties caused by digitally controlled machines and the feasibility of manufacturing the spindle. Contradictory constraints (concerning congestion and machining) are brought to light. It seems, however, that, once these constraints have been clearly expressed, the contradictions no longer really matter; it is always easy to reach a compromise. The problem lies in the fact that the contradictory constraints are not expressed clearly or at the same time. They are brought up by a large number of people who do not necessarily see one another.

Nevertheless, the deliberations in front of the computer are not a success. The method planner is not able to define a satisfactory solution. The question is given further consideration during visits to the axle manufacturing plant. The method planner wishes to see for himself exactly what the problems are. He asks the workshop technician in charge of the manufacturing line to join him in front of the machine. They talk things over beside the machine, using prints of the solid model. It becomes apparent that they define jigging points by analogy with previous solutions. They reason according to existing assemblies. They evaluate old assemblies in terms of their advantages and drawbacks, according to criteria which are not clearly defined. Their deductions are not at all based on the functional data or the theoretical rules to be found in trade manuals. These people are in the habit of reaching conclusions by comparing one case with another, even when they are assessing tolerance intervals. Procedures based on general rules or generic methods, such as functional analysis, are not compatible with their work habits. However, the introduction of these procedures does bear some fruit. The experimental work carried out on another axle, using these procedures, is partially re-used. In the same way, the documents drawn up during this experiment (a design conditions sheet and a machining plan) provoke elaborate discussions which testify to a mutual learning process (Hatchuel 1994).

Scene 5: A Changeable Cooperation Structure, Depending on Local Rules and Details

The game develops under the auspices of several people. Solutions consolidated in front of the machine tool and during repeated visits to the design department are called into question whenever someone new gets involved. When the forge controller goes to the design department, he consults the interface agent about the positioning of jigging points (figure 5).

He begins by talking about the static indeterminacy of the solution. This leads him to reconsider the positioning of jigging points and to ask further questions, such as "To which side of the die should forging defects be confined? The forge controller attempts to "optimize" the position of jigging points, but he does this on the basis of his own constraints (figure 6). Under these circumstances, what does the word 'optimize' mean? The forge controller tries to anticipate the impact of flaws in the forged product. He tries to reduce machining allowances. It is, however, surprising that this intention should derive from the forge, whereas, at first glance, it should concern above all the machine shop. In fact, this objective reflects the new strategy which the forge, previously considered as a mere supplier and threatened with externalization, is trying to develop to arouse the interest of its "customers."

Another situation provided an opportunity to decipher, to a certain extent, the ways in which planners act. They apply the same thought processes to component design and the definition of aids as they would

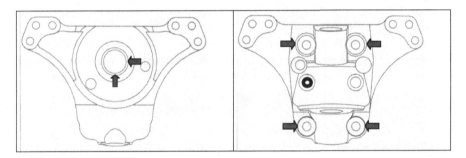

Figure 5
Jigging points as seen by the method planner.

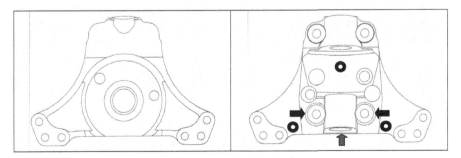

Figure 6
Jigging points as seen by the forge controller.

to the elaboration of machining plans. They create a number of situations, leaving themselves several alternatives to fall back on to. They take their final decision at as late a stage as possible (a way of doing things that can probably be attributed to the production environment and its constraints). The design department, in contrast, tries to define a single solution that will remain stable over time.

Thus, each person follows a different thought process, and each has his own constraints and methods. The design process presupposes that disparate thinking can converge on the same solution. Now, we have already seen that the general rules laid down in trade manuals, as well as recently introduced generic procedures, are not sufficient (or have not been up until now) to bring about this convergence. The latter only comes about after constant comings and goings between the people involved, and repeated adjustments. The new jigging point configuration, established with the forge controller, forces the machine shop to change its habits. The approach differs from that used for existing spindles, which the method planner took as a starting point. Such repeated adjustments generate a risk of introducing irreversible elements, depending on the order in which the different people intervene in the decision process.

Reasoning based on general rules remains insignificant for several reasons: limited distribution of new procedures, lack of adequate tools, unavoidable difficulties. The forge controller attempts to place jigging points at the same height on the machine part, thereby enabling the development of a base plane. In this case, he is acting according to a temporary, local positioning rule, linked to the software which drives the part control tool, which does not allow for point projections. Such a situation is common.

Compromises between professions emerge, above all, from specific and local practices. This appears to be more important than working on a set of general trade rules. The latter are already familiar to, and applied by, everyone. The difficulties encountered do not stem from this, but from local details which must dealt with constantly. Trade rules, which have already been assimilated by all who use them, do not present any obstacles and do not require compromise. Therefore, those who design new tools for computer-supported cooperative work (CSCW) should spend less time applying trade rules to them, in order to guarantee widespread distribution, than on concentrating on their cooperative agents. The latter should allow people to express temporary and local rules.

These local and temporary rules, which it is sometimes important to clarify, are often tacit. Thus, during a discussion with the interface agent, the forging method planner ends up explaining, without realizing it, one of the local rules that guide his thinking. It concerns the positioning of the ejectors (holes in the forging die that allow the part to be ejected mechanically after forging). The forging method planner intends to place them at the level of certain humps. It depends, he says, on their height. If they are too big, he will not be able to fit the ejectors into the die block, and he will have to find another place for them.

The same goes for the positioning of the blending radii on the forged product. The planner explains that there are not really any rules, except that the maximum radius must be used whenever possible and that the radius must never measure less than 3 mm. Once again, the planner has explained a rule without meaning to. This particular rule is very interesting, as it was entirely developed by the design department and the forge, according to the possibilities provided by the available software and to previous stamping experience. This tacit rule results from "cross-learning" that has occurred in the course of various past projects. It is, moreover, the product of a type of logic not known in design research. It is completely different from the logic used for the proposed new tools. It is a local creation whose fate cannot be established a priori.

Scene 6: Representation Work Becomes Autonomous

A designer complains that he does not have certain information on the spindle. He needs it to digitize the design. He demands, in particular, that the design department give him certain information which mainly relates to the assembly and machine shops. It is therefore absurd that this information should be provided by the design department. Besides, it seems that it is still too soon to carry out digitization. The other departments have not had enough time to do their work. Yet it is rather tricky to ask the designer to give up on his task, inasmuch as it represents the major part of his workload.

When project management was more sequential, the design department did as it pleased because it seldom consulted the other departments. Now the consultation process imposes a different work pattern. It requires, in particular, waiting for answers from other people. The impatience of the designer, who has begun his graphic construction, shows that the integration of knowledge is only just beginning.

Cooperative Agents

The type of graphic model used has a significant impact on the nature and depth of the discussions it engenders. Solid models and surface models[7] alone cannot stimulate remarks, questions, and suggestions. They must, at the very least, be accompanied by people who question, explain, assess, and enhance them, generally in an informal manner. This clash of opinions between part designers and manufacturers (from the forge and the machine and assembly shops) is not systematic. Yet it is in the manufacturers' interest to assert their point of view as soon as possible. A bad design is always likely to be impossible, difficult, tiresome, or expensive to produce in large quantities. Now, we have often noticed that their involvement is relatively feeble and irregular. We have seen that the distribution of graphical representations and observation sheets is not enough, and we have looked into the reasons behind this situation.

Scene 7: Expressing Oneself in a Virtual World

Graphical representations are stored in computer databases which the manufacturers can consult. An observation sheet is sent to them. They must sign it for approval and insert any remarks they may have. The sheet is returned, duly signed, without the slightest annotation. Yet the manufacturers often have something to say, and do so when they visit the design department. This is where difficulties and differing points of view are brought out into the open and new arrangements are made. Obviously, the difficulties brought up and the advice given by the manufacturing departments are not recorded on the observation sheet.

The inadequacy of the written document can be explained. In the first place, professional people find it difficult to express in words a problem they have detected in a drawing. It is often hard to give a clear name to a graphic element, whereas it can simply be pointed out if the person concerned by the remark is also present. In the same way, it is easier to explain manufacturing problems through visual gestures. Finally, the way in which a problem or a solution is expressed depends on how it is understood by the listener. In the presence of the listener, and with the drawing in front of us, we can adjust the amount of information we give, or repeat things differently. On the other hand, when we are alone in front of a sheet of paper, we are burdened with the question of knowing what to say without saying too much, but, at the same time, without omitting any information. Writing does not seem to have found a place in this world of machines[8] and drawings.

Besides, written documents tend to make the author's position irrevocable. This is due in part to the fact that recipients tend to consider written answers "firm and definitive." This tendency, apparent in the relationship between the different professions involved in the design process, is, moreover, reinforced by the contractual nature of the observation sheet, and by its role in design project management. It is a commitment. Now, as we have seen, manufacturers tend to prefer solutions that leave them room to maneuver and a way out. They give themselves the scope to face up to whatever difficulties they may encounter during production. Putting their position down on paper, without any possibility of looking into things further, is contrary to their methods.

Computerized graphic expression hardly seems more suitable. The graphical representations designed by the design department are available to the manufacturers in a computer database. Their authors, who work almost exclusively in this virtual universe, have equipped themselves with design software and use it to express their ideas relatively easily. Moreover, current software largely reflects the primacy of design over manufacturing. This computer language has become the designers' own language; it is not a shared language. Besides, exclusively geometric design scorns production constraints, as if a part could be designed on a universal level with no regard for the production procedures, machines, and specific know-how that will subsequently be used to develop it. The virtual universe belongs, first and foremost, to designers.

On the other hand, the forging and machining professions generally use physical elements (material, die, machine, etc.) as reference points. Virtual activities take second place. Therefore, they manipulate computer files and digital models with less ease than the design department. Besides, they are not preoccupied by the same things. Their concerns lie at the junction between certain features of the machine part represented, and the production tools that they use. Until now, graphical representations have scarcely provided any holds at all (Bessy and Chateauraynaud 1995) for people from the manufacturing profession. So these people prefer to point things out physically, and orally describe what they are demonstrating. It is up to the designer from the design department to express the significance and implications of this demonstration by means of a drawing.

Scene 8: Cooperative Agents at Work
The inadequate formalization of professional know-how, apart from general rules, is a major obstacle to cooperation between designers and manufacturers. Consequently, it is difficult for designers to take this know-how into account, especially with regard to local and temporary rules. Moreover, professional people have difficulty finding a hold on the graphic objects put to them. Although some models mean more to them

than others (the solid model, for example), their knowledge and constraints, when taken into account, are included implicitly in the drawing. If cooperation between designers and manufacturers is to be a success, each and every person has to get involved in interpretation and explanation. This does not make today's graphic design methods any easier.[9]

The interface agent developed the idea of adding symbols (small icons) to the solid model. Each symbol represents a hold on the object. This hold is specific to a given profession and has an impact on the design of the part. Looked at another way, the design of the part influences the hold. These new graphic objects give a meaning to the model that makes sense to the people involved on a local level. They make it easier for the different people to understand one another. They both support and formalize discussions. They could be jigging points, for example, or the places where part dimension checks will be performed.

The designers, having created an initial solid model incorporating the rules and constraints specific to each different profession, ask the manufacturers to come and work with them in the design department, in front of a computer, to improve the model. The interface agent and the method planner develop a provisional machining plan for the part, which is then used to determine the nature and position of the trade symbols to be added to the solid model. The same type of operation is performed with people from the other professions. Moreover, these people talk things over between themselves whenever their mutual constraints diverge. The formalization of their point of view via these symbols simplifies the development of a shared meaning and makes their discussions more fruitful. We have called these symbols *cooperative agents* (Laureillard et al. 1998). They are boundary objects (Star 1989) between the different professions, and mediating, go-between objects in the relationship between them. They reflect ways of acting and the restrictions on these actions. They make some peoples' decisions visible and, at the same time, provide holds for other people.

The introduction of this new graphic object modifies design activities. So, for example, the addition of jigging point symbols to the solid model shifts the emphasis of certain skills. A compromise reached by the forge, machine, and assembly shops and the design unit replaces the tandem of the designer and the machining planner.

Scene 9: Cooperative Agents Do Not All Have the Same Effect

Cooperative agents such as jigging points and check points effectively fulfill their mediating role. On the other hand, another agent of the same type seems to fail. It concerns the modeling of machining allowances[10] on the solid model. Apparently, the people involved do not latch onto it. Is not this agent relevant? In fact, those involved are not all of the same opinion regarding its importance in terms of saving on the machining process. It seems there are other priorities. There is another hypothetical explanation for this failure: perhaps this agent does not offer sufficient hold, because, for example, it incorporates too many constraints, such as the idea of a minimal chip (determined by the machine tool operator and dependent mainly on the jigging points) or the requirements of the forging profession with respect to drafts or blending radii and offsets. Should these two elements be split, thereby providing two distinct holds instead of only one?

This failure once again emphasizes the extent to which the cooperative agent concept is linked to people. It has meaning only inasmuch as it provides a hold upon which individuals can act. Before it is created, existing relationships must be closely analyzed, and the logic behind them clearly understood, as illustrated by the following example.

The interface agent and the machining method planner are assessing tolerance intervals. Together, they re-examine the machining plan. The method planner expresses its capability with regard to all the work measurements, and wonders which way the part will be facing during certain machining phases. He explains that there are several ways of fixing this direction: a temporary backing piece on each eye mounting, a V-block on the body of the kingpin, an automatic centralizing device on surfaces designed to support such a device.

Now, these means are not very accurate. Needing more precise direction, he asks the design department's permission to make a "localizing hole" in the part. The hole, which enables him to find his bearings on the part, represents a new hold, which may be considered as a cooperative agent. The creation of such an agent is, however, relevant only if it reflects a recurring situation in other design activities. Indeed, it is possible that the method planner's request is a one-time-only occurrence and that it is generally too difficult for him to foresee this type of requirement before the production of the first pilot lot.

The creation of this new cooperative agent is a result of the following actions on the part of the interface agent. By closely observing the way people interact, he identifies the requirement, then encourages the person concerned to say what he thinks and to put his needs into concrete form. He then tries to generalize the concept corresponding to this hold and this cooperative agent. He incorporates it into group work procedures, for example by adding it to the design specifications in the shape an explicit task. When presented to the rest of the people involved, it generates new ideas and remarks that lead to its improvement. It therefore illustrates what is needed in terms of holds to ensure that all the necessary adjustments are made.

Operational Summary

1. Graphical representations are closely linked to the organization of design work. They come into action throughout the design process.

2. These representations are used in a variety of ways, depending on the process phase and the people involved.

3. The very existence of a graphical representation is sometimes sufficient to get specific processes underway: beginning a design process without an official decision to do so, assessing solutions on the basis of the holds that different professions see in them, suggesting new ideas.

4. Manipulating graphical representations is an activity in which new ideas, questions, and knowledge come to the surface.

5. Graphical representations do not speak for themselves, even to professionals. They must be adjusted regularly so that they comply with the objects and representations that the people concerned are familiar with.

6. Graphical representations portray the intentions, methods, and constraints of those who designed them. Therefore, they reflect actions and processes already underway. It is possible to describe organizations and processes by observing graphical representations.

7. The methods used to produce and distribute graphical representations are a reflection of variable command-based relationships. When people decipher graphical representations, they are also deciphering, at the same time, the balance of power between themselves.

8. Graphic operations are accompanied by discussions, in the course of which many adjustments are made between the people involved and their different points of view. These discussions are often informal and spontaneous, and design decisions are made during them. Questions therefore arise as to whether these informal discussions should be managed and, if so, how.

9. Some objects (certain types of drawings, computer consoles, machine tools, symbols) seem to stimulate informal discussions more than others.

10. Rules are not all well known or completely formalized. People discover them during discussions.

11. Inter-professional cooperation develops through the re-design of some of these graphic objects.   The meaning of some of them becomes clear only in the context of discussions between professions.

12. Cooperative agents (inter-professional symbols) are not equally successful with the people concerned.   The fact that they exist implies that the procedures and relationships of the people concerned have been clearly understood and that generally applicable elements have been derived from them.

9

Rough Drafts: Revealing and Mediating Design
Éric Blanco

Rough sketches made by designers are instrumental in defining a new object's characteristics. They both reveal and influence the socio-cognitive design process. In the former role, rough sketches provide an account of the process and its periods of variable intensity and breaks. As mediators, they give a pattern to design activities. They draw attention to changes in the nature of the work and in discussions between the parties concerned (linguistic changes especially). They reflect, and are them-selves, representations of the product being designed. They are the prod-uct of this activity and an incomplete record of it. They are also resources which the people involved use to convince, explain, recollect, revise, imagine, come to an agreement, and so on. They therefore chronicle the activity, the relationships of those involved, and the way in which the future product is progressively represented. They provide insight into the design activity (Vinck et al. 1995). Therefore, they both reflect (as inaccurate disclosers) the socio-cognitive design process and act as mediating objects (active instruments).

Ethnographic observation and the recording of linguistic changes and graphic productions therefore help us better understand the engi-neer's work and the design process, which is generally thought to be cognitive, intellectual, and both rational and creative. Various ethno-graphic studies have already shown that the myth of the designer work-ing alone soon falls apart when processes of industrial design are investigated. (See chapter 4 above; see also Bucciarelli 1994.) On the contrary, the work is shared by various individuals, who contribute knowledge and methods specific to their worlds. As a result of these studies, certain design tools and the way they were used in industry were called into question (Blanco et al. 1996). These studies called for more effective use of coordination mechanisms. Still, it was better to pursue these investigations with a searching examination of the interactions between designers. The cooperative design environment referred to in this chapter was instigated and set up with this aim in mind.

A Cooperative Design Experiment

This study's reference situation[1] is a design experiment involving five people located in the same room for a limited, pre-determined length of time. In two three-hour periods, these people have to design a product, from drawing up a rough set of specifications to establishing manufacturing plans. The participants are neither designers nor specialists in the product concerned. They are chosen for their skills in the areas related to their supposed role, in compliance with the shared design theory. These roles are defined according to the following three criteria:

• The functional module, made up of two people, makes sure that customer requirements are met. This is the only module to have detailed knowledge of these requirements. It drew up the specifications and gave them out to the rest of the group a few days beforehand. This module speaks for the customer during the experiment.

• The structural module's purpose is to endow the product with the best technical performance capacities possible.

• The manufacturing module, made up of two people, must make optimal use of the production means placed at its disposal.

The product to be designed is a machining assembly for shaping wooden parts for the furniture industry. It is an innovative product corresponding to a pre-identified requirement. However, those involved do not know the solution. At first glance, the design of this product does not require lengthy calculation. It is therefore suitable for this experiment.

The elements drawn from this experiment include a video recording and twenty intermediate objects developed by the group during the design process. All related discussions were then entirely re-transcribed using the videogram. The intermediary objects consist of sheets of A3 paper. When analyzed, these elements (discussions and intermediary objects) were supposed to shed light on both the cognitive processes involved and the collective and socio-technical means employed to find a solution. This analysis is based on the theory that intermediary objects play a mediating role in design activities (Vinck et al. 1995; Jeantet 1998). These objects are used to track activity and the process's temporality. They represent both the intentions and skills of the people involved and the future product. Nevertheless, to understand the mediating role played by objects, they must be observed in action and, in this case, in relation to the discussions between the people involved.

First I am going to show how these rough drafts, or intermediary design objects, mark out the process's specific temporality. I will show that when objects are changed, the group dynamism is interrupted and a new phase begins. I will then show how they can be used to track the emergence of the product and the progressive construction of both the problem and the solution. Then we will try to understand their mediating role in the socio-cognitive process. The people involved use these objects as conventional aids, enabling them to create a common system of reference on which to base future activity. Finally, we will think about their role as temporal mediators. We will therefore consider the relative stability of these objects as memory backups. We will discover that this function is much more limited than we had imagined.

Objects That Disclose and Compose the Process's Temporality

During the process, various objects were created by the different people present. These objects are a record of the design activity and provide a glimpse of its temporal structure. They are considered active as long as they are at the center of the workspace and as long as they have not been put aside. The recording of these objects, of the moment when they emerge, and of the periods during which they are actively used in discussions, provides the first overall view of the project's progress. The temporal diagram illustrating references to the various objects (figure 1) shows that, in 6 hours, 20 objects were produced and used. They can be divided into the following five formal categories:

- texts and diagrams made up of words organized into lists, diagrams, and tables

- sketches (simplistic schematic representations of the product)

- overall drawings, which represent the mechanism as a whole and which follow the conventions of engineering drawing with regard to hatching, representation of axes, etc.

- definition drawings, each representing a single element and respecting the rules of engineering drawing

- models (either three-dimensional physical objects or settings of objects that illustrate how a product is fitted together).

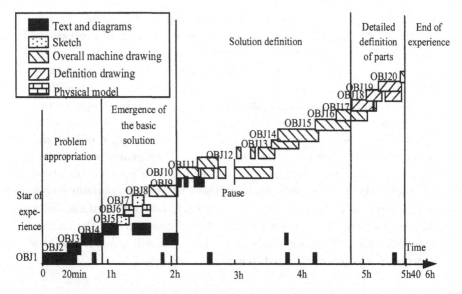

Figure 1
Chronogram of references to various objects.

On the whole, the objects tend to emerge one after the other and to replace one another. They succeed one another rather than clutter up the discussion space. There is little overlapping, except when one object is used to create another. Objects that have been abandoned are seldom returned to. The only exceptions are object 1 (the specifications, figure 2) and object 3 (the list of functions that the product must perform, figure 3).

The type of formalism used is organized temporally; it is not random. Therefore, texts and diagrams are used mainly at the beginning of the process, followed by sketches and models, then overall drawings, and finally detailed drawings. The formalism therefore changes progressively as the design develops.

Objects That Trigger the Start of a New Phase
Some objects correspond to a break point; they bring about a transition from one distinct phase of the activity to another. The boundary is not situated between two objects; it is represented by a single object. The objects that succeed it differ from those that preceded it. This object is the boundary. We can therefore assume that it triggers the start of a new phase. The appearance of object 4 is an example of this. Object 4 is a table (figure 4). For the first time, the product is defined in terms of a solution. This marks the disappearance of the functional representation,

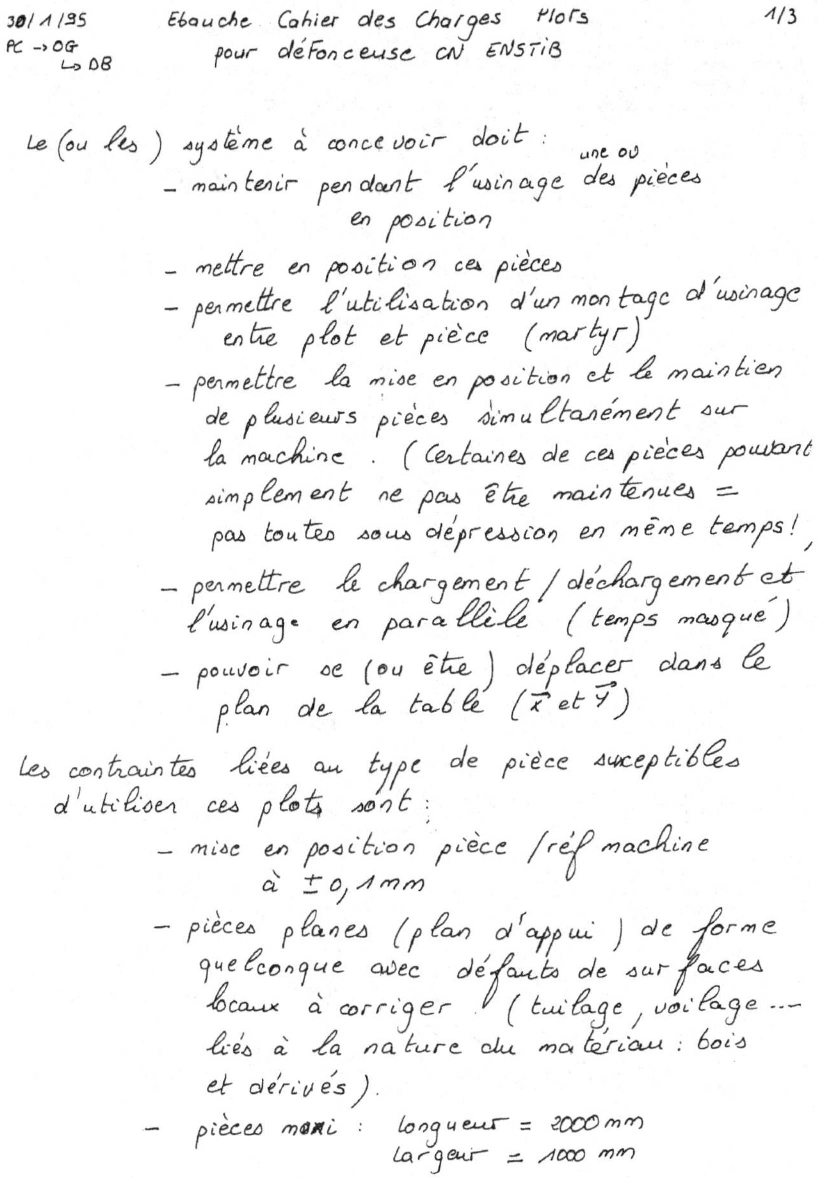

30/1/95   Ébauche Cahier des Charges   Plots   1/3
PC → OG   pour défonceuse CN ENSTIB
    ↳ DB

Le (ou les) système à concevoir doit :
                                                    une ou
        – maintenir pendant l'usinage des pièces
                en position
        – mettre en position ces pièces
        – permettre l'utilisation d'un montage d'usinage
            entre plot et pièce (martyr)
        – permettre la mise en position et le maintien
            de plusieurs pièces simultanément sur
            la machine . (Certaines de ces pièces pouvant
            simplement ne pas être maintenues =
            pas toutes sous dépression en même temps!)
        – permettre le chargement / déchargement et
            l'usinage en parallèle (temps masqué)
        – pouvoir se (ou être) déplacer dans le
            plan de la table ($\vec{x}$ et $\vec{y}$)

Les contraintes liées au type de pièce susceptibles
    d'utiliser ces plots sont :
        – mise en position pièce / réf machine
            à ± 0,1mm
        – pièces planes (plan d'appui) de forme
            quelconque avec défauts de surfaces
            locaux à corriger (tuilage, voilage ...
            liés à la nature du matériau : bois
            et dérivés).
        – pièces maxi : longueur = 2000 mm
                         largeur = 1000 mm

Figure 2
Object 1: specifications.

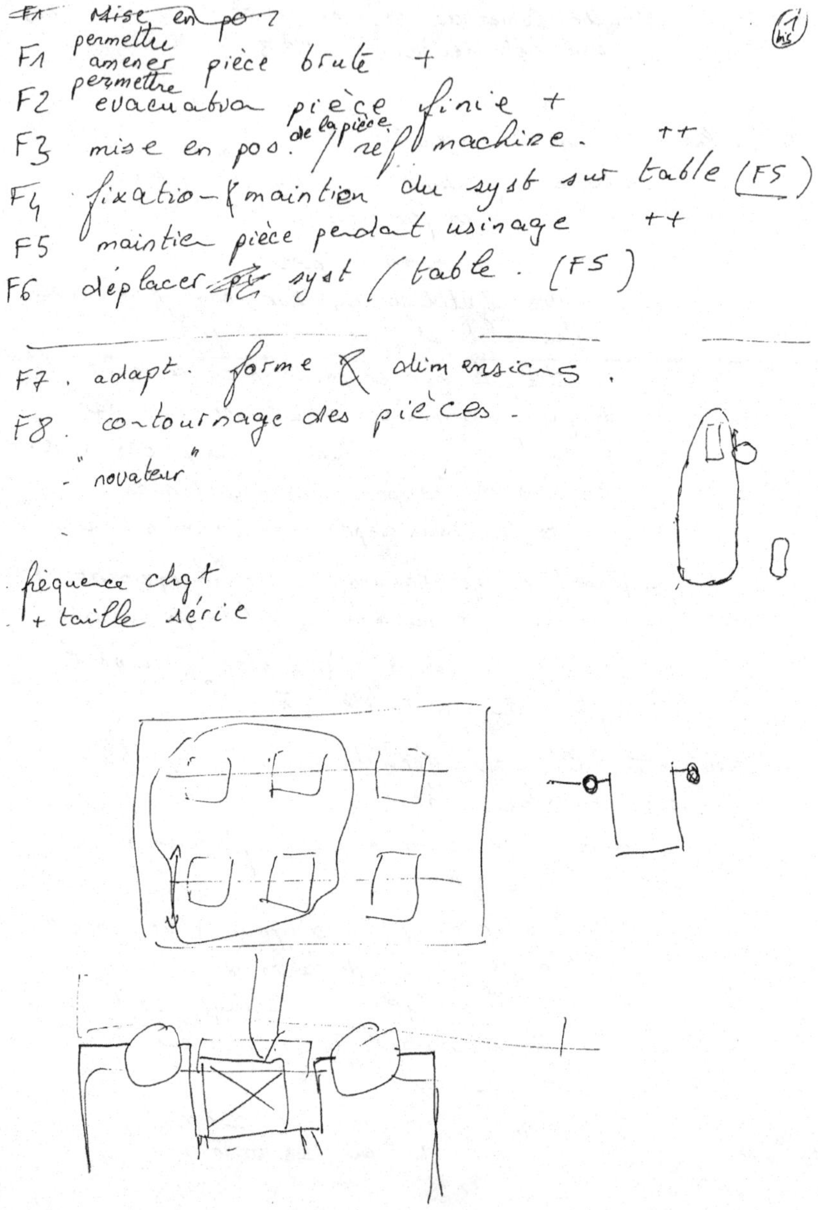

F̶1̶ Mise en po?
F1 permettre amener pièce brute +
F2 permettre evacuation pièce finie +
F3 mise en pos. de la pièce / ref machine.   ++
F4 fixation-{maintien du syst sur table (FS)
F5 maintien pièce pendant usinage   ++
F6 déplacer ̶p̶c̶ syst / table. (FS)

F7. adapt. forme & dimensions.
F8. contournage des pièces.
     "novateur"

fréquence chgt
+ taille série

Figure 3
Object 3: functions list.

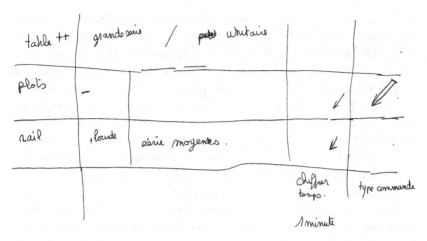

Figure 4
Object 4: solutions table.

transformed by one of the members of the group into a range of possible solutions. Several basic solutions are therefore put forward and organized in the form of a table, which is a way of systematically organizing and structuring text. This represents a change in the representation language.

These break points make the process irreversible; their corresponding objects embody the decisions that have been made, in part implicitly, by the group.

Objects That Mark Design Phases

Once the nature of the objects has been analyzed (a process that involves a detailed examination of their content), further dissection helps identify phases and break points. Moreover, these objects should not be dissociated from the actions, discussions, even silences[2] to which they are related (a designer's personal or private creation, the presentation and explanation of this creation to other designers, debates and discussions about the object, the use of the object as a basis for another drawing, or a single designer's thoughts on an object). The phases thus identified correspond, on the whole, to those described in writings on design methodology (Pahl and Beltz 1996). Four phases have therefore been defined.

Problem Appropriation

The first 50 minutes are given over to analyzing requirements and functions. The objects produced and handled are mainly texts: specifications, a diagram of the main functions and a list of functions. The specifications are supposed to provide a clear picture of what the requirements are. Yet it takes 20 minutes of talk, in the form of questions fired at the "function" person who drew up the specifications, to reach a clear understanding of what is being asked. The "function" person speaking for the customer then says: "Right, shall we get down to some work, if you haven't got any more questions?" So begins the next phase, during which a diagram illustrating the functional analysis is supposed to be produced. Object 3 results from this phase, during which the designers appropriate the problem and haggle over the list of functions that the product must perform. A common understanding of the problem emerges from these discussions. The graphic objects, as well as providing a basis for these discussions, progressively record the conclusions reached. Objects 1 and 3 are the only ones to be used again in the next phases.

Establishment of a Basic Solution
The second phase begins with the creation of object 4, a table. In this case, the "structure" person suggests three types of solution. During this phase, four other objects will be developed: two sketches illustrating two basic solutions (objects 5 and 7) and a model (object 6) made out of plastic cups and a book. Throughout this entire model-creation period, the designers will use plastic cups to illustrate one of the basic solutions. They will use this rudimentary model to simulate the product's various operating phases. With object 8 (figure 5), a basic solution commands attention and is settled upon. We then move on to the formalism of engineering drawing. It is during this solution-development phase that the individuals involved use the widest variety of formalisms. The other phases are, on the contrary, dominated by a largely predominant type of formalism. The solution-development phase lasts 80 minutes, 20 of them spent on object 8 alone.

Defining the Solution
This phase is the longest, lasting 2 hours and 35 minutes. It develops the basic solution fixed upon at the end of the second phase. Thirteen objects are created. These consist mainly of roughly sketched overall plans based on the rules of engineering drawing. The "structure" person does most of these drawings, while discussions between the various people involved focus on these graphic objects. Little by little, figures concerning dimensions and other calculations appear on the drawings, as well as some sketches when the basic solution is reconsidered. Object

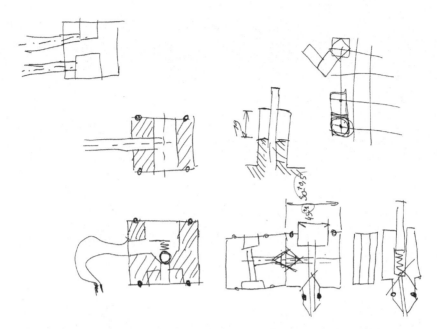

Figure 5
Object 8: first overall plan.

9 stands out from the rest in that it is developed by the "function" person and is kept apart from the others.

Detailed Definition of Parts
During this phase, detailed drawings that will be used to manufacture the product are created. The first drawing in the series (figure 6) is done jointly by the "structure" and "manufacturing" people, keeping manufacturing requirements in mind. During this phase, four objects are developed: three detailed plans of the various parts and a drawing in perspective. The "manufacturing" person labels one of these objects "piston" in accordance with a verbal agreement reached by the group.

The analysis process clearly illustrates the role that objects play as temporal markers of the design process. The phases that they define correspond to dominant trends and not to strictly specific activities. Several objects are used in different phases. On a single object, we find marks corresponding to activities specific to different phases (for example, a sketch on a detailed drawing). Therefore, although from a general point of view it is obvious that the organization of activities follows a particular

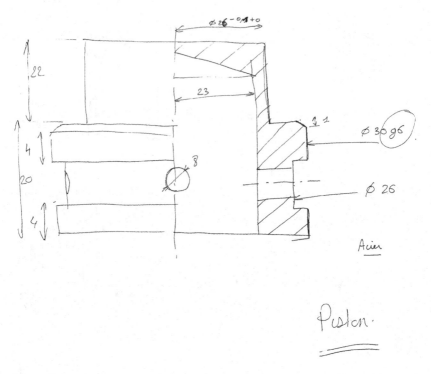

Figure 6
Object 17: first detailed drawing.

trend, the process's non-linear elements should not be underestimated. We can therefore see how intermediary graphic constructions record the emergence of a product and the progressive construction of both the problem and the solution.

Objects as Conventional Aids

In this part, we will try to understand the mediating role that intermediary objects play in the collective socio-cognitive process. Norman (1993) talks about cognitive artifacts[3] and emphasizes that the cognition between the artifact and its user is ordained. These artifacts may be memory aids, such as checklists, or action aids. Norman shows that, by relieving people of a certain number of tasks, cognitive artifacts help to structure their users' activities.

In our case, the cooperation among several designers revolves around these objects. Thus, the design process is shared out between several human beings and the intermediary objects that they produce and use. Therefore, the latter are socio-cognitive artifacts with a role in the design process.

The people present represent dissimilar points of view (function, structure, and manufacturing) that are differentiated a priori by their goals and their language. During the process, they both confront one another and try to cooperate. The explanations of each person's point of view, of conflict, of the building of a common meaning, and of active cooperation are all based on the artifacts used. By analyzing these artifacts in relation to discussions, we can observe the comprehension mechanisms in play between group members.

These analyses reveal the existence of several coordination modes, all used by the designers. They reveal the existence and development of joint conventions, for example those related to engineering drawing or those with a common meaning emerge during the process. From this point of view, the intermediary object operates as a conventional aid[4] (Dodier 1992) in these coordination modes. The purpose of analysis is therefore to trace transitions from one to the other by relating in detail the interactions around these intermediary objects.

Transition from a Cultural Conventional Aid to a Pragmatic Aid

The experiment begins with a clarification of the rough specifications by the "function" person who established them. This second reading is the first phase of the experiment. The "function" person expounds the problem and the customer's requirements, already partly converted into functions in the specifications. This person leads the discussion; he also does nearly 80 percent of the talking during this period. At this stage, the problem is presented in terms of functions; only a few references are made to solutions. The "function" person considers the specifications, distributed a few days beforehand, to be a solid aid that will give everyone clear insight into the problem. When the experiment begins, he suggests reading and discussing them quickly. In his opinion, this should take 3 minutes. The following transcript reproduces the dialogue. (Here and below, FUN1 and FUN2 are "function" persons; STR is a "structure" person; MAN is a "manufacturing" person. The meaning of the bold type will be explained shortly.)

{0'00"}

FUN1:   We are going to try to design overall machining supports or assemblies for a machine used to machine er parts made out of wood or materials derived from wood. So er we've defined first of all, I've managed to draw up specifications, we've defined a certain number of functions, **I'll try to give you a quick idea of what I've written down.** . . .

STR: **Er, how long is this going to take?**

FUN1: **Three minutes, in three minutes** it'll be done and we'll get down to work. I'll explain the specifications I gave you, OK? Right, so it's . . . these assemblies er the assembly, right, a certain number of functions have been defined, so hold one or several parts in position during machining so the part or parts it's simply because we have a relatively large work table we can machine several parts at . . . at the same time on the machine, so we want to be able to machine several parts at the same time on the machine, position these parts so that we can use a machining assembly between . . . er the corner support and the part, what we call a martyr, allowing us to place several parts on the machine at the same time.

MAN: **I don't understand what a martyr is.**

FUN1: A martyr is a part that we insert between the assembly, therefore the object we are going to design today, and er the part itself.

MAN: **So it's extra, right?**

FUN1: Extra yeah so the fact that we can er machine er several parts er at the same time or in series er on the same machine so with er if possible er the fact that some parts on the table on the machining assembly are at a given time either fixed or not fixed to make assembly and disassembly easier er in the meantime er allow loading-unloading and ma . . . and machining at the same time for [inaudible] reasons. We machine in one place, we load a part in another place and we unload a third.

MAN: **Doesn't the table move?**

{2'35"}

In fact, reading object 1 will take more than 20 minutes. Why? Our theory is that the specifications imply that a certain number of conventions are common to all the people involved. They are not. The way they are written and the vocabulary used presuppose that the group is familiar with the environment surrounding the technical system described and its functions. Now, the participants do not have all the information needed to interpret the contents. From the beginning of the explanation, some of the vocabulary used is not understood. These conventions, whether they are local or cultural, correspond to specific knowledge (relating in this case to the machining of wood). Therefore, object 1 does not act as the conventional aid that its author supposed it would. The individuals involved must appropriate it and create their own representation of the future system's environment. Object 1 therefore represents an inter-subjective (pragmatic) type of coordination that involves the establishment of new conventions enabling the group to use this object as a starting point for their work. This transition from one coordination mode to another is visible in the questions asked, the statements made, and the wording used by the "structure" and "manufacturing" people during the "function" person's presentation (set in boldface here). Throughout this period, the comments made by the different members of the group are all requests for explanations.

FUN1: Yeah . . . er so 3 or 4 cm minus . . . so 277 minus er well the height of the parts they're panels even though I said that the maximum height was 170 millimeters. Eh, right, these are exceptions, OK? So 170 millimeters, that leaves us with 100 millimeters. The minimum tool's 90. That means that in theory the support measures 10 millimeters OK? That's . . . [shows STR a sheet of paper] . . . that's here, eh? So, these constraints, that means that to use these assemblies we're going to . . . we're going to there are some borderline cases we won't be able to use.

STR: Hold on.

MAN: **I don't get it.**

STR: **Me neither.**

FUN1: We've got . . . between the pin and the table we've got 277 millimeters.

STR: Yes.

FUN2: Hmm.

FUN1: The maximum part height, maximum height, is 170 millimeters.

MAN: **So, this 170 is the 40 max you pointed out here.**

FUN1: Yeah . . . hang on . . . in gen . . . yeah but in general er . . . the average must be closer to 40 millimeters than 170. . . . If I take the maximum limits we've got 107 millimeters minus the tool, 90 if we take the worst case.

STR: No.

FUN2: No, take the best case.

STR: Take 170 millimeters . . . that must be the worst case.

FUN1: OK, I agree, yeah but . . . otherwise it's going to be. . . . So we are in agreement but that means we can't use a tool without a support OK [laughter] so you agree this case is irrelevant.

FUN2: So we're going to have to [inaudible]

FUN1: So we agree then that that means we have to plan on as thin a machining assembly as possible.

MAN: **So what's that got to do with 18?**

FUN1: OK, considering that we are going to exclude a certain number of thick parts . . .

MAN: **The connection with 18 then . . . I understood that the support can measure more than 18.**

FUN1: Er well 18 er it's to do with having . . . we've got forming tools to shape the parts . . .

MAN: Hm.

FUN1: And that means we have to be able to move the tool down 18 mm below the bottom of the part.

MAN: So into the table if we don't use the support.

FUN1: So into the table if we don't use the support . . . the indications given here are critical, sure, in some cases er it doesn't work . . . that simply means, I repeat, that we must have the thinnest machining assemblies possible, so we do have a little bit of choice.

STR: **Can you just tell me one thing?**

FUN1: Yeah, no, no, we have got a little bit of choice . . . considering that we can take the table to pieces, we can take it to pieces since it's quite deep and we can gain 5 or 6 centimeters like that. . . .

The Object's Tacit Conventions

The members of the group suppose that the object is comprehensible because they believe that they have enough common knowledge, methods and references to understand one another and the object. They presume that the object is based on tacit conventions that need not explained. Thus, many of the features of object 1 are not discussed. They are inherent in the group's common conventions. The "technical culture" that all the participants have in common should, a priori, enable them to get the job done without redefining all the terms used. For example, the diagrams used in object 1 are not clarified. The "function" person assumes that everyone is capable of deciphering them.

But these conventions are not necessarily familiar to everyone in the group. It should not be assumed a priori that those involved are capable of deciphering the diagrams. However, it is assumed that they are, as none of them requests an explanation.

Moreover, a single object can be used in several coordination modes. The preceding extract provides an example of how several types of coordination can operate at the same time. In this case, it is not the diagram that is unclear but the reasoning used by the "function" person to estimate the height of the technical system to be designed. He does not explain how he has simplified things, especially the fact that he has taken the average height of wooden panels produced on this type of machine as a starting point.

When the group perceives a lack of "solidity" in the conventional reference, such as the above-mentioned failure to understand, it momentarily switches from one coordination mode to another to fill the gap between the differing interpretations. Nevertheless, this re-adjustment may come at a late stage. The following extract shows how the word 'part' is misunderstood. Re-adjustment occurs only after 109 minutes of discussion. Before this, the word 'part' is used about 260 times.

{approximately 1'49"}

STR: It says so in your specifications

MAN: [inaudible]

FUN1:   OK, er it's . . .

MAN: No because one meter out of two we won't [inaudible]

FUN1, to STR: It's five . . . ah no, I didn't say five or six supports, I said five or six parts. It's completely different.

FUN2:   Yeah, and . . . how many parts per support?

MAN: **Eh? What's a part? Ah, the wooden part . . .**

FUN2:   Well yes, the wooden part.

MAN: OK, yeah, now we haven't got the right vocabulary.

FUN2 [reading a piece of paper]:   A maximum of five to six parts can be on the machine.

FUN1, to MAN: It doesn't bother me if one part is fixed to four supports.

MAN: So. . . .

Words, diagrams, and drawings are not unequivocal representations. They give rise to changes in the coordination mode. An object's ability to be a conventional reference should not, therefore, be taken for granted. If the object is to act as a coordination aid, the group must progressively create a shared knowledge basis.

The Role of Rough Drafts in the Creation of New Aids

We have already seen that certain intermediary objects accompany each change of phase. The people involved use them as a basis for switching from one procedure to another. As has been pointed out, these boundary objects trigger the start of a new phase. Now let us look at what happens to these objects.

Object 4 triggers the change from the first phase to the second. After lengthy discussions in which the group gets a grasp of the problem, the "structure" person orally presents three types of solution that, according to him, provide a response to the problem as it was expounded during the first phase. He then draws up a table (object 4) with three headings representing the three possible solutions: corner supports, table, rail. These terms are used in expressions such as "table" (for the present system), "We need to think about how to improve things," "corner supports" (for "mobile aspiration systems"), and "rails" (for "much more rigid systems"). These refer to the discussions that took place during the 5 minutes preceding the creation of this object. They are not defined. It is assumed that, as a result of these discussions, they are perfectly clear.

These three words that appear in object 4 are used to designate the three solutions throughout this entire period, whereas these solutions constantly change. The table stabilizes the vocabulary. The group then uses the object by pointing to the solutions referred to and by employing demonstratives:

STR: Should we try to develop **that solution?**

MAN: Well, in my opinion, with **that one** we're going to be much too . . . expensive and complicated.

FUN1: Yeah, it's going to be complicated, yeah.

STR: Yeah.

MAN: Never mind, we'll develop **that one** and we'll discover we're too expensive.

STR: Yeah, yeah.

Object 4 reflects the "structure" person's current view of the design problem. It represents a change in the way the product is represented. It is the first object that presents the product in terms of a solution, rather than in terms of functions or requirements. Therefore, all the activity over the next 10 minutes will be based on this table, and one of the three solutions presented in the table will be chosen. Object 4 therefore helps to stabilize the design problem.

However, the problem is not only stabilized; it is also transformed (represented). It is presented in terms of a range of solutions rather than in terms of functions to be performed and requirements to be met. The problem has become three solutions. The group's initial objective was to respond to the requirements laid out in the specifications. It has now become "Choose from these three solutions." No other type of solution will be proposed by the group in spite of two attempts to broach the subject. The first is made by the "structure" person before the creation of the table: "These are the main possibilities that I see, I don't know if you see any others." The second is made by the "function" person once the table has been created:

(57'36")

FUN1: Are you sure there are no other solutions to what we are trying to do?

STR: Right now I can't think of any.

FUN1: You don't have any others for now.

STR: Well, we can try to find some but . . .
FUN1: No, but it's. . . .

The members of the group never begin to look for other solutions. All their discussions are now based on the table. This limits the range of possibilities.

The table operates in another way in the collective cognitive process. It puts the terms 'table', 'corner support', and 'rail' on equal footing, as if the three solutions were comparable or interchangeable. Now, these terms sum up different problems and decision types. 'Table' corresponds to a minimal design choice consisting in improving a solution that already exists. 'Corner support' is a more generic term used in the specifications' title. We could, moreover, invent solutions such as an "improved table" with "corner supports" fitted onto "rails." However, the table separates these solutions and presents them as alternatives rather than ingredients. In any case, once the table has been drawn up, the group will treat these terms as identifying alternative solutions. A decision process, based on object 4, will then single out one of these three types of solution. The object really does support and contribute to the new perception of the product. It becomes a conventional aid of local importance in the design process that follows. It does not imply that those involved all attach the same meaning to it, but simply that this meaning will never be re-negotiated.

The conventional aid shows that it is not necessary to create a common language to enable coordination between the different members of the group. On the other hand, it is a good idea to use the wide variety of viewpoints to create objects that provide a framework to discussions. Nevertheless, we must bear in mind that the relative solidity[5] and the limited relevance of these conventional aids make the process irreversible. This irreversibility probably stems less from the object itself than from the fact that it was created by a group and that it has been acknowledged and become a reference for each person. Taken out of this collective work context, it loses a lot of its meaning. We could go so far as to say that, if this irreversibility stemmed from the physical object only, it would be easy to cancel it simply by changing the object. That's not how things are. We cannot disregard its presence and its indelible role in producing a meaning and fixing it in memory.

If we compare this situation to others in which objects are used to convey and transfer information, one characteristic of the rough draft emerges: the fact that they are not highly codified makes them an excellent tool for integrating knowledge and different viewpoints into group action. Furthermore, their ability to preserve and transmit a meaning turns out to be poor. On the contrary, highly codified intermediary objects are based on high-level conventions. Such is the case for the definition plans provided to the planning or manufacturing department. Unlike sketches, they do not leave the manufacturer much opportunity to influence the design. They are closed objects (Vinck and Jeantet 1995; Mer et al. 1995) that operate on a prescriptive basis. The sketch and the rough draft, although they make some use of high-level conventions (the basic rules of engineering drawing) give the manufacturer a chance to influence the design only if they are created jointly. They operate on the basis of pragmatic cooperation.

Objects as the Failing Memory of the Process

Let us now return to the role that objects play as temporal mediators and investigate their relative stability as memory aids. What is the role of these intermediary objects over the course of the design process?

There is very little overlapping. Few objects are handled, looked at, or discussed simultaneously. The production and use of objects is relatively sequential. It is difficult to say exactly what is behind the transition from one object to another (for instance, the removal of one sheet of A3 paper and the introduction of another). Some are full, others much less so. Some objects are homogeneous and support one type of task only. Objects are rarely retrieved once they have been removed from the work space. Their service life lasts from approximately 10 minutes (for object 7) to nearly 90 minutes (for object 10).

Object 10 is, in this respect, in an unusual position. Its service life is much longer, which means that it is carried over from the first part of the experiment to the second. The designers use it to kick off the second day. It is used a reference in the creation of object 12 and is only cast aside when object 13 is developed. It consists of various groups of diagrams corresponding to different actions on the product. The shape of the corner supports is, for example, anticipated on it.

Action Mediation vs. Memory Mediation
This instability of sketches is interesting analytically because it casts doubt on their usefulness as memory aids in the design process. Object 4 is, once again, a good example. It is put aside after around 1h09'45" when object 5 is created, then picked up again by the "function" person after around 1h25' when the group finds itself in a fix. The group has decided

that the "corner support" solution is the most appropriate but has run into problems regarding the localization of the corner supports in the machine's coordinates. The "function" person therefore draws the group's attention back to the "rail" solution: "Could we do it with this or not?" This transition from one solution to the other is accompanied by a 18 seconds of silence, during which the "function" person goes to get the table (object 4). He studies it but does not seem able to find what he is looking for.

This table acted as a backup and a mediator in the choice of the "corner support" solution. The group cannot, therefore, use it to justify reconsidering the reasons for their choice. Taken out of its original context, it does not provide the information required to go back on the decision. It is difficult to see any meaning in the few words and arrows that appear in the table, even for one of the people who designed it and only a few dozen minutes after it was designed. Therefore, the efficiency of these objects as memory aids is not necessarily proportional to their role in the process. Object 4, which was a crucial element in the emergence of the solution, cannot be used to reconsider the reasons behind the decision made. In this respect, an irreversible process is set in motion. The object appears not to have the same role when it is an action aid as when it is a memory aid.

Although the temporality introduced by objects is different from that created by discussions, it rapidly reaches its limits. Another incident that occurred during the experiment leads to the same conclusions with regard to a different object. Object 10 (3h10'45") reveals an element on the "corner supports" which is identified by the designers as an "insert." Its purpose is to compensate for the possible stress caused by installing the wooden panels, to avoid damage to rapidly made joints. This insert disappears in the final solution. At present, its usefulness is being debated. The designers have to make a considerable effort to justify its presence, in spite of the backup provided by the drawings. This brings us back to the remarks that Grosjean and Lacoste (1998) make when studying shift changes in different hospital departments. Instructions are passed on via objects such as tables and instruction sheets. In spite of these documents, the oral transmission of information is fundamental and complementary. In the same way, in our design situation, we can see how written objects are used by those involved to back up their actions, but also that they are relatively inappropriate as memory aids. The designers themselves are unable to use them to justify their choices.

Previously Distinct Actions

Objects lack depth. When the group retrieves an object created at an earlier stage, during a temporal process involving distinct actions (additions, erasures, and reformulation), it is incapable of determining the importance of the various elements. It fails, for example, to distinguish something that has been decided upon from something that appears on the object but has not been collectively validated. Everything that appears on the retrieved object is equally important (unless precise explanations are provided). Such was not the case when the object was in use.

The importance of an inscription is determined by meta-scriptural or linguistic information. It is often oral interactions that define the importance of these inscriptions. For example, the fact that 'piston' is underlined twice in object 17 means that the "manufacturing" person has validated it before going on to define other elements. On object 10, the person representing the structural module circles figure 15, stipulating that it is to be checked. However, after the event, only individual recollection and the collective reconstitution of the circumstances in which an inscription was made enable the group to recall what its purpose was (double underlining, rings, arrows, etc.). In the same way, elements are selected and assessed in a specific, momentary context. It takes more than diagrams to put all the information together again.

Intermediary objects therefore provide a record of the design process, but they fail to keep the memory of it alive. The irreversible process thus created is reinforced by the fact that these objects provide no insight whatsoever into the decision process that they supported and coordinated. The transformation/complementation (Mer 1998) that they represent developed during discussions. Although at the start they often represent someone's mental picture of the product, they are subsequently shared, modified, annotated, crossed out, validated, or rejected by the group. They bear the mark of all these discussions, but they no longer enable those involved to recall them distinctly. Time freezes them like snapshots. Insight into the processes that created them would make it possible to reconsider decisions.

Unlayered Sedimentation

The design process speeds up with the introduction of object 8. It is created by the "structure" person to put an end to a conflict with the "function" person. This conflict occurs after a misunderstanding between these two individuals. In the discussions that take place before the creation of object 8, a new solution is suggested; it consists in separating the holding-in-position function from the putting-into-position function. As it happens, one of the problems raised by the "corner support" solution was how to position the supports in the machine's coordinates. This separation creates two elements: aspiration supports (which hold things in position(and stop supports (which position the wooden panel in the machine's coordinates). The "function" person also talks about separation, but from a different point of view: he wants to dissociate the fixing of the support onto the table from the fixing of the panel onto the support. The misunderstanding turns into a conflict when the designers refuse to budge:

STR: You're really getting on my nerves.

FUN1: Because earlier well . . . I don't know or you didn't understand but . . . I'll explain it again, we said that if we put, if we place the aspi . . . the aspiration system at the same time we're . . . we're talking about an aspiration system if I put the part in when the support isn't fixed to the table . . .

STR: Yeah.

FUN1: If I put my stop part in . . .

STR: Yeah.

FUN1: Once the aspiration system has been set up, it might move, that's what we said earlier.

STR: That's what you said, yeah.

FUN1: **Yes, but hold on, I'm the boss, I'm the customer, aren't I?**

[laughter]

FUN2: **Now you're changing hats.**

STR: **True.**

FUN1: Don't you agree? We did say that, didn't we?

Object 8 will clear up this conflict by illustrating the differences between the two points of view. The first two sketches of object 8 (figures 7 and 8) show how the "structure" person sees the solution. They bring the roots of the conflict to light: the two people are not talking about the same thing.

The "function" person can use this sketch to reformulate his request. The "structure" person responds to this by creating two other sketches (figures 9 and 10). These sketches respond to the "function" person's concerns.

The "manufacturing" person then takes the problem up from a different angle by assessing the suggested solution. He considers that the

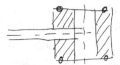

Figure 7
Sketch 8-A.

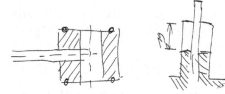

Figure 8
Sketch 8-B.

presence of pipes on the machine tool table is not reliable or satisfactory. His rejection of this solution gives rise to a new solution a few moments later: rapid connections that pass beneath the support (figures 11 and 12).

The group has not made any decisions regarding the maintenance supports and the stop supports. The suggestion made by the "structure" person will be implicitly adopted. Therefore, a new solution emerges. In 8 minutes, the stop support has been separated from the table while maintaining its position in the machine's coordinates. The question of whether to disunite the aspiration system is solved at the same time as the rapid connections hold the supports in place. The "structure" person then puts forward a much more general view of the mechanism (figure 13). This overhead view shows where the rapid connections are situated on the table and leads the group to consider the possibility of joining the maintenance supports in pairs on the side of the table. He then completes one of the diagrams by adding a support connected to the side. In the end, object 8 represents the chosen solution in its entirety. The finishing touches will be put to it during the third phase of the experiment (figure 14).

The step-by-step analysis of the development of object 8 clearly illustrates how the sketches and rough drafts used during the design process create conflicts between differing points of view and make cooperation

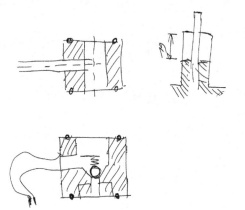

Figure 9
Sketch 8-C.

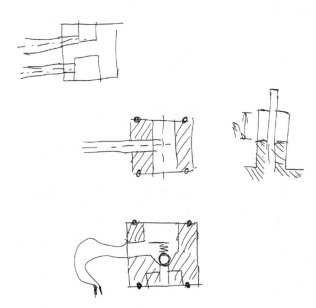

Figure 10
Sketch 8-D.

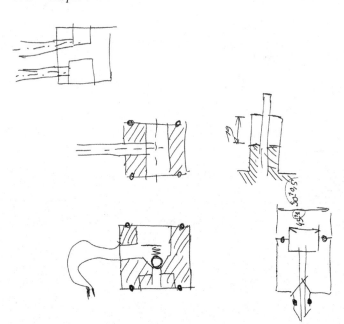

Figure 11
Sketch 8-E.

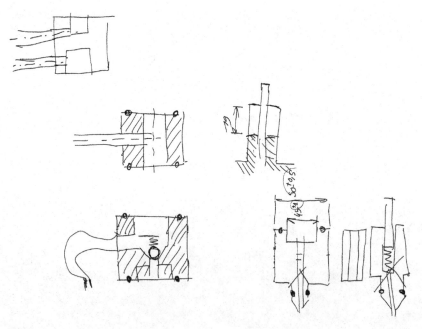

Figure 12
Sketch 8-F.

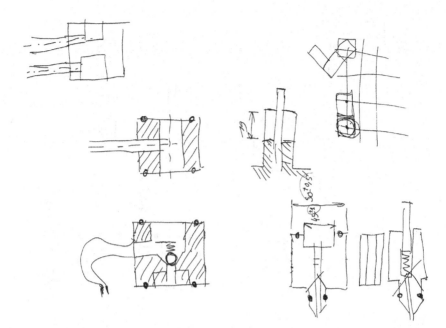

Figure 13
Sketch 8-G.

possible: a solution gradually takes shape as mutual representations are considered. A new object emerges as each person's "mental picture," and the interactions between the group members, are put down on paper. Of course, all these inscriptions are not of equal importance. However, a posteriori, there is nothing in the sketch to suggest this.

Operational Summary

1. The attention paid to the intermediary objects produced and used by the group broadens our understanding of design processes.

2. When the cognitive process involves several people, it develops mainly through verbal and graphic interactions.

3. The traces that it leaves behind are mediators in the process before becoming a chronicle of it. They are used, along with linguistic analysis and the diachronic analysis of action, to qualify the phases of the process and some cooperative cognition mechanisms.

4. One of the cooperative cognition mechanisms leads to the development of ruptures and irreversible elements.

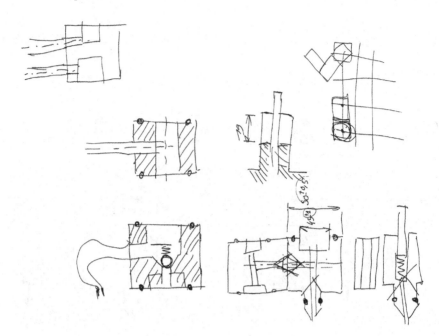

Figure 14
Sketch 8-H.

5. Relative irreversibility can be explained by the inability of these traces to stimulate recollection of the process leading up to it.   Therefore, paradoxically, the support's physical stability fails to jog the collective memory, because it puts productions that were not originally of equal importance on the same level.

6. On the other hand, the physical support's mediating role was confirmed during the process: it objectified events and/or represented individual mental pictures.   The result was the collective construction of a future object through a series of representations that succeeded and replaced one another.

Epilogue: Approaches to the Ethnography of Technologies
Dominique Vinck

Technical developments have left a considerable mark on modern society. The concepts of progress and change are associated with those of technological innovation and increased knowledge. The elements that make up our daily universe are redefined, in part, in industrial companies and research laboratories. Moreover, there is no shortage of discussion about current changes and their consequences. There are also plenty of scientific and technical writings presenting the mechanisms and laws of the phenomena involved in the most advanced technologies. Finally, there is no lack of methodological recommendations and standards that are supposed to define correct procedures for engineers and technicians.

Now, the technical practices developed in companies, offices and research laboratories are still very poorly understood. In spite of speculations on the causes and social consequences of technology, and the proliferation of written methodological prescriptions, it is still very difficult to get an accurate idea of what really goes on. Design and innovation activities and technology are dealt with in an abstract manner that makes them invisible (Orr 1996). In fact, we still know very little about actual design practices and the use of technology. Since these practices are now at the center of almost all professional, domestic, and leisure activities, we might as well say that we know very little about what people really do.

Besides, theories on technology and industrial reality are usually dispersed among several disciplines, some of which ("engineering sciences") are interested in objects, techniques, and methods and some of which (human and social sciences) concentrate on the social fabric. The former have shown very little interest in the people involved and their activities; the latter have often neglected the many material objects involved in technology. Until recently, technology was considerably discredited, reduced to a material level which was, supposedly, of no interest to the sociologist, or to a mere instrument that only had importance in terms of results and effects.

Today, human sciences are witnessing a revival of interest in the study of real practices, including the objects used by the people concerned. Industrial activity has become a relevant research subject again, and what goes on in design offices is starting to draw the attention of researchers from both the social sciences and the engineering sciences. The former are finally daring to go into design offices looking for an answer to the question "What are societies composed of, how do they reproduce or change by manufacturing or using objects, by acting on the matter and the people around them?" (Latour and Lemonier 1994, p. 10) And researchers in the engineering sciences have come to see that the designing of new tools is contingent on a better understanding of how people behave at work.

The present work fits into this new anthropology of technologies. It aims to provide a better understanding of our society through ethnography. This epilogue attempts to size up this new ethnography, which aims to explain not only human actions, but also the objects and performances used by each of the parties concerned.

Performance Ethnography

The present ethnography involves making investigations and writing reports. It combines in situ observations, open or semi-directive interviews, records (lists, counts, etc.), and the observer's participation in certain activities ("participant observation"). This method of inquiry requires the observer to stay at the investigation site for a long time in order to fully understand its value, its main features, and its subtleties. The observer records what he sees, hears, and experiences in a journal or a field book. He writes down conversations, keeps the documents he has received or produced (photographs, screen prints, computer files), and compiles data. These personal souvenirs are also a sort of record that the observer will use later to draw up a report meant, in particular, for people who are unfamiliar with the environment under study. Developing and controlling a form of writing used specifically for explaining the situations studied is just as much a part of the job as managing investigations.

This is not a new practice. Elements of it can be found in numerous works. Ethno-archaeologists, ergonomists, sociologists, and engineer all make detours into the empirical observation of human organizations and the objects they manipulate. Ethnologists use this method to describe the material basis of cultures in an endeavor to understand their symbolic system. Ergonomists describe the behavior of operators, either to gauge their discomfort and improve their working conditions or to establish reference situations that can be used to develop new working conditions. Ethno-archaeologists observe how living societies take possession of certain techniques, whereas archaeologists only present the marks left by these techniques (and thereby recreate objects and operations from societies that have vanished). Nevertheless, the work of the aforementioned disciplines is hardly satisfactory for the study of technical practices. The first reason for this has to do with the way in which objects are built; the second has to do with the environments studied.

Material Practices and Mediations

The construction of the object of investigation depends on the authors' assignment. Although engineers sometimes conduct long and thorough studies, they seldom publish detailed reports on them. On the contrary, more often than not, they only keep hold of a few general technical models or principles. Sociologists do not act any differently. The only elements retained from their observations are those that allow them to study the perceptions and values that guide technical actors, their standards of behavior, their social identities, and their power, apprenticeship, and distinction patterns. Anthropologists use their observations to elucidate, for example, the underlying structure of the culture they are studying.

This tendency to neglect material mediations can even be found in works by authors who have intentions similar to ours, including the authors of various articles published in Moisdon 1997. Like us, those authors pay close attention to tools (in their case, management tools), and they take account of how those tools are designed, implemented, and used. However, they mainly analyze only intentions, methods, principles, and groups of actors. When describing a graph, they discuss only the logic behind it, as if its materiality were irrelevant. Part of the graph's depth and opacity is therefore left unanalyzed. Thus, a lot of information is omitted from many technical analyses. We attempt, in contrast, to consider the various types of mediation, including material mediation, just as Akrich (1992b), Weil (1999), and Chapel (1997) did. We believed in the value of going into the details of the actions and objects we observed and explaining their concreteness.[1] Thus, we dealt with practices and the opacity of the objects handled,[2] the details of situations, their socio-technical fabric and the meandering route taken by people. Our aim was to understand the inseparably social and technical "structure" of situations, people and actions. Our work consisted of analyzing the effective practices and mediations employed on a local level—practices and mediations that make technology and society intricate and distinct.

A Study of Today's Ordinary Technical Practices

Accurate and detailed technical reports have been published for a long time, but they mainly concern "traditional" or "exotic societies." They discuss, for example, the working of clay in India, Mexican potters, the domestication of the pig, the use of the lasso in Finland, and irrigation and the measurement of time in Tunisia.

Since the 1980s, other anthropologists have been studying contemporary Western technology. Some of them have studied technologies that have now become banal: the use of telephones (Akrich 1992a,b) or video recorders, or the maintenance of photocopiers (Orr 1996). Others, including Latour and Woolgar (1979) and Traweek (1988), have concentrated on the temples of modern knowledge, or have focused on the transfer of technology (Akrich 1995).

The industrial world has not escaped the notice of the social sciences, but the production department has received most of the attention (Freyssinet 1992). Ethnography of technology is seldom employed in modern companies. Most documentary evidence consists of studies of technical heritage or of research into dying technologies. In the case of such studies, ethnography provides a record of industrial history and worker know-how (Tornatore 1991).

A few studies concentrate on advanced modern technologies. Downey (1992) decodes the culture related to the development of computer-aided design. Forsythe (1993) observes specialists in artificial intelligence. Woolgar (1985) analyzes the significance of computer technologies. Scardigli (1992) and Gras et al. (1989) study the design of Airbus cockpits, computing, and reproductive biology to discover how social actors give purpose to technical developments. Their approach, however, differs from ours. These researchers are interested, above all, in culture, values, and perceptions, as well as the symbolic systems and power structures inherent in the development and implementation of these advanced technologies. From their anthropological point of view, the main thing is to understand the modern world from the spontaneous perspective of the actors. Anthropology thus attempts to re-construct the dynamic of meaning in today's social and technical world. Ethnography is a step toward understanding this identity. For these modern-day technical anthropologists, ethnographic research should concentrate on looking for the meaning of the world from the point of view of its inhabitants (Hess 1992, p. 3). Chapter 5 of the present volume also has this end in view when it explores the reasoning processes engraved in the designer's memory and in the objects around him.

However, our project aims to differ from the previous approaches. On the one hand, it studies the technical work done within modern industry without focusing on either advanced technologies or worker know-how. The goal is to investigate the everyday work done in design offices or by the engineer in the field. Very little research has been done in this area. (Bucciarelli 1994 and Button 2000 are exceptions.) We also try to explain performance (i.e., what is actually produced through human activity), not only the meaning of the world from the point of view of its actors.

A Question of Performance Rather Than of Meaning

Anthropology studies cultures from the point of view of their symbolic system, among other things. It examines, for example, the relationships between technical practices, the representations underlying them, and the social patterns involved. It takes objects and gestures into account because they have meaning—because they remind people of who they are, how they are supposed to act, and what they are worth. Thus, the anthropologist studies the specific characteristics that hold meaning for the members of the culture being studied. Taking this as a starting point, he explains belief systems, culinary rites, collective organization, architecture, the content of songs, etc.[3]

When it is transposed to the world of the industrial design office, this approach helps us to determine how designers and draftspeople give a meaning to speeches, practices, and objects. Their technical activities are contingent on words and images, which they interpret according to an assortment of implicit relationships that it is our job to reveal. The words used to think and act technically are sometimes the same (for example, "open" or "closed") as those used to talk about an object, a person, or a group. Although individuals do not necessarily realize it, these expressions are used to think about material things as well as social issues. Both of these areas are made up of joint, interacting "symbolic materials." The notion of mobilizing scales, discussed in chapter 4, can apply to both people and work tools. Technical objects and activity are, in this respect, socially significant. It is just a question of decoding their symbolic and operational syntax.

Nevertheless, our project aims to do more than merely search for a meaning. It also attempts to explain performance—i.e., what is actually produced through human activity. Our theory is that performance, aside from technical or functional elements that can be explained by natural sciences or mathematics, is contingent on human behavior and language. Performance, like meaning, is something that ethnography can help us to understand.

Until the 1980s, anthropology was based on the assumption that an object or a gesture has a material dimension and a symbolic dimension. The former explained the object's purpose and the limits it imposed. The study of the object fell within the scope of the exact sciences. The anthropologist was concerned only with the symbolic dimension. The ethnographic techniques employed, even when they involved a detailed description of behavior, were only a transitory step toward studying the only thing that counted: the meaning.

The ethnographic approach adopted in this work differs from that described above. It focuses on analyzing performance and refuses to deal with it only from the point of view of the natural scientist. Ethnography is, in this case, used as a means of understanding action and its consequences, namely the production of socio-technical hybrids. It implies interest in practices, in groups (the world of design and that of manufacturing), and in objects (a drawing, software, waste). It supposes that a detailed report will be composed of all sorts of mediations.

Our project has therefore consisted of explaining the nature and the content of a given result (statement, tool, collective identity) and giving an account of how it was created. Our theory is that the identities of the people involved—who is a technician, who is competent, who is a designer, who is a good research worker, who is the most creative, what is so-and-so's specialty or experience, and so on—results from negotiations. All these questions are debated within the group. Individual identities are therefore created during and by action. Those involved in a situation ascribe interests and influences to one another and reach agreements concerning methods and the results obtained. In accordance with Garfinkel's (1967) precepts, an agreement among various people is accounted for by describing the procedures used to reach this agreement. This explanation is not technical, nor is it based on general causes (the distinction stratagem, power relations). It has its roots in the present, local situation. The elements relevant to the explanation are the actions and words of the people involved, in the specific context in which they find themselves.[4]

Nevertheless, our ethnographic àpproach differs from ethnomethodology in the way the object is dealt with. Ethnomethodology does not really take the object into account. It is assumed that it exists, even if the user does not do anything with it. The person involved is not supposed to have any philosophical doubts regarding the existence and nature of things. As a result, the object and its action are not included in their analysis; the only things that count are the coherence of action, the order of things, and the procedures used to produce this order. We are therefore obliged to complete the work of anthropologists and ethnomethodologists. The former were only interested in the symbolic dimension of facts; the latter concentrated on procedures and coherence from a human point of view. In both cases, the material aspect of activities was neglected, and explaining it was left to the exact sciences. Our project has been to re-introduce it into the analysis process.

Ethnography of the Social World of Technology

The technical universe is a social universe that is less widely known than that of the Peul shepherds or the Jivaro Indians. Designers in (for example) the automobile industry or the electrical industry belong to a culture not known to the "man in the street" or to the social science researcher. Although they are members of our society, they work in places that seem exotic to us. And yet they design the objects that we encounter (trucks) or touch (electric switches) every day. They are also the magicians we call when misfortune strikes—when a road accident occurs or a photocopier breaks down. Getting to know them better means getting to know our own society better. This is the primary purpose of technical ethnography: to decode the identity and singularity of these fragments that make up the modern world.[5]

The Foundations of Modern Society

Our collective identity and the objects that characterize our culture are fashioned in design offices and in industrial companies. Studying them is tantamount to studying one of the sources of our civilization.

A priori, it is reasonable to believe that these shrines of science and technology are emblems of Western rationality. Western culture, once close to nature and dominated by religious beliefs, seems to have left the "pre-logical" era behind and moved on to an age of "logic" characterized by scientific knowledge.[6] Thus, studying scientific laboratories and design offices comes down to studying the most characteristic features of our society. Such evolutionist theories have, nevertheless, been strongly criticized by anthropologists. The history of societies is not linear. Even science and technology have experienced ruptures and revolutions in the course of their history. The theory of continuous progress is being challenged. It implies that there is only one reference by which to measure the different situations observed. For example, from an anthropomorphic point of view, it is questionable whether typing with two fingers on a computer keyboard is a form of human action that is superior to writing beautiful and ornamental letters with a quill. Moreover, in-depth studies of distant societies by archaeologists have revealed that these societies possessed excellent knowledge and command of their own sociotechnical complexity. Even a recent study of animal societies (Latour and Strum 1991) forces us to re-examine our ideas on the beginnings of human evolution: baboons are much more accomplished and skilled in controlling frenzied social complexity than we had imagined, and, in view of chimpanzees' systematic use of tools, we can no longer associate tools only with humans.

Conversely, ethnographic studies (Latour and Woolgar 1979; Lynch 1985; Vinck 1992) have put laboratories and modern technological innovation in a much more "uncivilized" and "primitive" light.

Technical ethnography therefore leaves old clichés behind to analyze these shrines of Western society more closely. The plan is no longer what it was in the 1970s (to prove that these shrines are more ordinary and less rational than they appear). The aim now is to understand the logic processes specific to these shrines, and to identify the ways in which they help people to bring their projects, and ours, to successful conclusions.

The World of Technology Is Not Closed

The aim is therefore to better understand some of the mechanisms at play in this fragment of our society involved in technical design. Our attention is drawn first to the inaccessibility and the integration of the society being studied.

Anthropology, having cast evolutionist theories aside, concentrates on the singularity of each culture. Some authors treat each society as an isolated entity, forming a functionally integrated whole.[7] Others see the individual society as a group of elements borrowed from other societies but coherent on a local level. The anthropologist's work therefore involves going into the field and drawing up a monographic and holistic report. Nowadays, the possibility of isolating cultures in order to study them without any outside influence is being fundamentally challenged. The same holds true for a company or one of its departments. For example, a design office cannot be studied as a closed space, cut off from the rest. On the contrary, we have shown[8] that people, texts, tools, and theories from other places pass through these departments. The actors take these elements and re-configure them. They produce new documents, objects and incorporated skills that are connected to various networks and put into circulation. The technical worlds studied from an ethnographic point of view in this work are open; boundaries are never clearly defined. Besides, they rarely match pre-defined organizational specifications.

Questions can also be raised about how the elements contained in the social spaces studied are incorporated. We can split these elements up into categories (scientific, technical, organizational, socio-economic, political, cultural, etc.), just as anthropologists have done for other societies.[9] This is what Thill (1973) and Traweek (1988) did for the field of high-energy physics.[10]

Distinguishing different dimensions implies that observations, which are not necessarily empirically linked, are selected and placed in the same category (political, scientific, organizational). If we proceed in this manner, we are forced to consider how these dimensions interact with one another within a functional whole. On the contrary, as today's anthropologists postulate, we should be studying the local interactions that reflect the transversal nature of the object being studied. Using existing documents as a basis for creating new ones (chapter 7) and using a recently developed part to meet a new requirement (chapter 5) are concrete examples of how objects become are crossing dimensions. By studying these examples and describing how objects are adapted and coordinated, anthropologists attempt to portray the composition of a given place. Functionality, or the convergence of all these elements toward a single end purpose, is never presupposed.

The Codes and Implicit Conventions of Technical Work

Our type of ethnography tries to keep track of another dimension to technical activity: the dimension of tacit conventions, explicit and implicit rules, and codes and languages that allow technicians to work together and understand one another. By analyzing words and actions (pointing, drawing a line) in a design situation, we are both revealing a number of implicit codes and showing how the parties concerned develop them, use them, play with them, forget them, and rediscover them. Our project differs from structural anthropology[11] in that, rather than comparing design offices to bring general designer grammar to light, it examines our respective locations and objects by comparing observations.

People have been studying the unspoken codes behind technical practices for a long time. Interesting examples can be found in Kuhn's (1962) work on paradigms and disciplinary matrices and in Wittgenstein's (1958) work on tacit agreements and the unspoken rules behind linguistic stratagems (drawing a graph, proving something mathematically, aggregating data). In the field of technology, Constant (1980) describes the socio-cognitive models that underlie and govern the aeronautical engineer's activities. The technological paradigm that Constant puts forward defines the sort of enigma that engineers think they must solve and the type of approach that is generally assumed to be valid.

The Treatment of Objects

Our ethnographic project takes account of actions performed on a local level, with and through intermediary objects. The physical dimension of these objects is also taken into consideration. Our theory is that the object cannot be considered from the social point of view alone, neither can it be reduced to a dual technical and social causality. Any explanation must therefore take the physical presence of these objects into account, without losing them in a web of sociological causality and naturalist or structural analyses. It apprehends technology by restoring its depth (or weight—see Kaufmann 1992) and its interactional contingency.

The Materiality of Things

Technology can be defined as an object because it is thought of as rooted in matter and physically active. The engineer therefore believes it is objective, even if he is simultaneously aware of its contingency, which is connected to the way it was developed, knocked together, adjusted or set. To put an end to explanations based on social constructivism and re-establish the unavoidable presence of the object and its performance, he puts forward supposedly irrefutable arguments based on the existence of tangible proof and the fact that "it works."

The objectivity of technology is not self-evident

Nevertheless, the objectivity of technology cannot, by any means, be taken for granted. One need only read the accounts given in this work to realize that the actors are continuously working on making this objectivity visible through drawings, calculations and prototypes. They develop it, view it, produce it, and maintain it. Otherwise, it would just be a cumbersome, meaningless leftover.

We have already seen that the tangible presence of technical objects implies that they are visible and active. The shielding disk, the functional dimensioning and tolerancing software, and the refuse bins mentioned in chapters 1–3 are very good examples of this. Visibility and action presuppose a large number of actors: demonstrators, popularizers, salespeople, legislators, fitters, maintenance workers, publicists, trainers, controllers, adjusters, experts (to do the explaining), decision makers (to allocate the resources). The texts, sketches, tables, screwdrivers, and rags used to bring these objects to life and to make them efficient and visible should also be added to this list of actors. All this is necessary, although the object gives the impression that it commands attention all by itself.

The object's physical materiality could, in fact, be blinding, especially when it works well from a technical point of view. We thus tend to attribute its power to its materiality (its nature, structure or intrinsic logic). On the other hand, when its power fails or when we are bothered by its cumbersomeness or its physical disorder, we tend to see only the human dimension: the people that did not do what was expected of them, inadequate training, organizational rules and structures that do not work properly, prejudice, excessive technocracy, and so on. When it works, this tends to mask the way in which it exists and acts.

However, the sociologist should not get things wrong. He is used to denouncing this sort of blindness. He refuses to believe that the strength of technology is self-evident. He shows how it depends on the "soft" social world hidden behind technology. He denounces the apparent stability, power and rationality that emanate from technology. He shows that these things hide the true nature of technology. He replaces the machine and its supposed intrinsic qualities with a multitude of people, who are the only valid reasons behind the existence and the power of technology. He talks about organizations, industrial relations, markets, professions and trades, worker know-how, power patterns, how efficiency evaluations are made, or the underlying structures of culture and symbolic systems. In doing this, he destroys the opacity of objects and relates them to sociological causes. He sees only social mechanisms. We have tried to avoid this sort of sociological reductionism.

Technology holds out against being relegated to a technical and social fact
The analysis of technical objects is balanced between the denunciation of underlying sociological mechanisms and the recognition of intrinsic

technical efficiency. Technology, when related to its social conditions, slips away from us. It resists its sociological dissolution and compels recognition through its materiality. (See chapter 3.) But when the observer tries to apprehend this materiality, he is immediately referred to multitudes of human beings and texts.[12]

When we try to apprehend the sociality of technology, we are reminded of the consistence of objects and our perception of the object as a social fact is threatened. Now, the people hidden behind this technical object (adjusters or researchers, for example) vouch for the existence and efficiency of the object that we are trying to explain from a sociological point of view. The operator perspires on the job; the engineer explains how a machine works to a visitor; the user guide reminds us that there is an after-sales service; a part's design bears the name of its creator. Nevertheless, although the technical object and its performance are vouched for by so many people, they remain difficult to apprehend. They are fugitives. They avoid both sociological dissolution and their supposed objective existence.

The difficulty therefore lies in establishing the reality of an object through the crowd of human beings and intermediary objects related to it: plans, project texts, specifications, technical files, brochures, test reports, expert valuation results, technical notations, lists of results, economic appraisals, etc. Without these things, technology would not be accessible to us. However, they hide as much as they reveal. Engineering drawings reveal an object and at the same time deform it. Technical ethnography runs into two problems here. On the one hand, it is still unfamiliar with this host of intermediaries, objects, texts and human beings that obscure our view of modern technology. On the other hand, it has difficulties in seeing technology through all these elements.

Sometimes technology seems to have been almost stripped bare, with the mass of intermediaries having disappeared. The operators, technical guides, and salespeople have been gone for a long time. In this case, the technology is mute, as it often is in a museum. It is dead; it no longer tells us anything. In contrast, when it is alive in the form of a highly efficient machine, it is surrounded by a host of intermediaries that put it on show.[13] There are two sides to our work, then: on the one hand, we observe and analyze the intermediaries or mediators (Hennion 1993) that make technological action visible; on the other hand, we get behind the screen that they form. This problem is all the more real since technology (as in the case of CAD and drawing, for example) has been less hardened by tangible material elements.

Technology in the act and making

Technology, drawing, and procedures make sense only when they are in action and in use. Technology is ephemeral and ever-changing.[14] It exists only because we use it, but at the same time it depends on the use that is actually made of it. When it works, a crowd of intermediaries takes over to make it tangible, but the arrival of this crowd changes the very nature of the technology in question. (See the description of functional dimensioning software chapter 2.) Each use of the technology makes it exist in a different way. It is thus a question of seizing each occurrence.

Some technologies (e.g., the bins described in chapter 3) are linked to very present physical objects that dominate the surrounding intermediaries. Their materiality is thus an important mediator of their presence. However, it is in danger of screening other mediators responsible for their performance: explanations and user guides, sanctions for incorrect use, and so on. The materiality of an object is one of the mediators of presence and technical performance, but is by no means the only one.

Studying objects while they are being designed or used plays a major role in our ethnographic approach. The idea is to get to grips with the process by which the active presence of the numerous intermediaries (both humans and objects) is synthesized, then to transform this into specific performance. Here, "performance" means what is produced in whatever register is used: technico-economic productivity, technical demonstration of the unsuspected possibilities of a machine or product, operator virtuosity (Dodier 1995), beauty of the machine's movement, turnaround in power relations, identity of a professional group, exalting destructive power, or demonstration of the potential violence of a technology (with weapons, for example). The ethnographic project thus consists of understanding the way in which technicians, specifiers, and users bring to light recognized and shared performances.[15]

Technology as Seized by Mediators at Work

Our ethnography tries to follow and report on the various mediators who, through their involvement in the technology, help us to understand it. To begin with, this means following the designers, operators, adjusters, trainers, and others. If we wanted to extend the study, it would also mean following the ergonomists, sociologists, economists, and others, as each in turn provides a new route of access to the technical object as it undergoes all its changes. This approach would allow us to see how these mediators depend on one another to bring the technology and its performance to life. To this list of mediators many others can be added:

schools, companies, distribution networks, media, political movements, legislators, and government representatives (for example, for environmental protection and occupational safety). They help to form the technical act through various channels, including denunciation, validation, improvement, moralization, or erudition. Approached this way, the technical act is a collective production (Hutchins 1995).

The importance of such mediators can be measured in the case of flight simulators used to train pilots. The engineers dig into the far reaches of their imagination to produce lifelike flight simulators with built-in storms, fog, and relations with other members of the crew. Nevertheless, these simulators are unrealistic; so many environmental mediators are missing: passengers, heavy airport circulation and air traffic, competition among airlines, the never-ending search for profitability.[16]

In addition to human mediators, there are material mediators, notably intermediary objects. Technical performance is generated, transmitted, observed and managed through such objects, whose importance is often underestimated: user guides, strings, bits of sticky tape, agendas, modeling dough, and so on. These objects certainly are present and have roles to play, but not necessarily in the way one might expect. They must therefore be marked, followed, and their action and opacity understood by moving from one to another, according to the relations set up between them during the action. Hence, reporting on the collective production of technology leads into a vast research program. It should pave the way toward an anthropology for our technical society (Bijker 1995).

Technology Understood in the Course of Action

Following the course of action in this way requires close proximity with the action. In fact, it is a question of understanding performances, stratagems, versions, tests, series, and uses, not a question of principles or ideas about objects. As can be seen with respect to prototypes (chapter 9), and as shown by Dodier (1995) with respect to technical virtuosity, performance occurs face-to-face between operations and technical objects, within the framework of a local scene, backed up by its own measuring instruments and commentators, and dependent on the performance event. In the case of design processes, the designer looks from one plan to another and from a draft to the computer screen, and sometimes he exchanges looks with a neighboring designer. He also thinks about the object by miming its operation and checking its representation: "Tap, tap. Seems OK there." The designer's "intellectual" work is also physical (Béguin 1997) and collective (Poitou 1997).

Performance (the object that comes out of the machine or the production level achieved during the day) is the outcome of a collective action. Owing to the high number of causes for variability, performance is different with each outcome. Each time there is a reproduction of performance the work that goes into it is different as the conditions of the production change: the machine is more worn or heated; the operator does the job better and better; the structure learns. (See chapters 3 and 7.) The identity and properties of the technical objects, just like the identity and solidarity built up in groups and trades, are the results of these actions. They are called into question with each reproduction. Each technical performance calls on the capacities of the machines, individuals, and groups. (See chapters 3, 4, and 6.)

One cannot understand objects without looking at the actions and actors involved in their making. Now, these same objects are approached differently according to the actors. The actors develop contrasting attitudes (Dodier 1995). At one end of the scale the object is respected. The actor treats the whole of the object with care or considers what it is or has become over time.[17] (See chapter 6.) At the other end of the scale, the object is brutalized. The actor prefers to adopt an attitude of human domination over a thing, striving to reveal all its secrets, to find out what it is capable of ("Let's see what you've got in you"), or to impose certain rules on it. (See chapters 4 and 5.) When it commands respect, by showing its performance or being silent (if broken down, for instance), the object transcends the fascinated group's hopes. When a brutal attitude is adopted, the object is reduced to an excuse or a simple means of enhancing the identity of an individual identity ("We've really got someone here") or that of a groups ("The X workshop has managed to keep to its deadlines").

During the action, the actors continually change their attitudes, from technical performance hypostatized as the cause of their fascination to the painstaking task of adjustment and negotiation where performance is only an ephemeral effect. In the first case, the technology is idealized: its indescribable power either fascinates or frightens. In the second, it is treated in an offhand manner and reduced, for example in the case of a competitor's technology, to customs, conventions, specific interests, and even the scheming of its designers. It can also be branded as a conniving illusion, a salesman's gadget, or just blind hard-headedness. These relations with the technology (hypostasis vs. denunciation) are set up in varying ways along with the action to guide it, justify it, or call it into question.

Following the Mediations

Technology in action is a synthesis or a node in a network of fragile links: instruments, user guides, plans, etc. Each of these intermediaries can be mobilized in two ways: either the technical drawing or specifications are taken as read, sealed by the medium on which they lie, and followed accordingly, or they are caught up in the movement to which they testify. This is why, when a drawing or a prototype moves from one workshop to another (chapter 8), the protagonists are confronted with different definitions of the object and the situation. It is then a question of knowing whether to faithfully transmit the object defined (at the request of the customer or the product to be brought out) via a series of things (drawings, databases, etc.), or whether it is above all a case of continuing the movement of transmission and mobilizing constantly reworked and transformed objects and actors (Hennion 1993). The transcendence of the technical object faced with the immanence of relations depends on this series of confrontations.

Following design intermediaries thus involves being less interested in the causes of efficiency (technical validity, economic profitability, social representations, etc.) than in the procedures launched. It is a case of seeing how things move around and get changed. It is a question of qualifying, on the one hand, the efforts needed to express and define the relations set up between various elements using material and human media and, on the other hand, the work necessary to put these media into circulation, reinterpret them and allocate them. The observer must follow these movements and get into contact with all those who make selections or act on the movement of beings. It is a question of seeing how they are understood or rejected, adopted, and adapted.

For the intermediary circulating from one office to another to set up a relation, and for the technology in action to develop a performance or the designer to be recognized as such, users must also keep their eyes and ears open and their hands ready to co-produce performance. This supposes that the users understand the language spoken to them (i.e., the language of industrial design or object handling). It also supposes that their expectations have been partially put in place by other intermediaries who prepare the user for this action: an engineer he trusts, a machine or a label that he recognizes, a covering signature.

Confidence in a plan, a procedure, a colleague, or a machine has to be built up; it does not exist in new situations. Actors believe in the beings they wish to believe in. They elect a certain number of these beings and rely on reference points and "handholds" (Bessy and Chateauraynaud 1992, 1995) to guide them in their actions and judgments. A fine analysis of these movements helps to reveal the various apprehensions generated during technical actions in brief moments of uncertainty, before their properties become fully evident. After this moment, during which there is hesitation in judgments and the way the action should be started, there is a distinct difference between the performance of the object and its recognition by a subject. Sometimes, depending on their own trajectory, actors refuse to give the technology credit; they therefore reduce it to an unbearable set of random elements, an empty carcass or a meaningless movement. An outside observer, therefore, as a person who does not understand the technical action being developed or the drawing he is looking at, sees only a lot of agitation, metallic noise, or scratching out. In this case, the technical action and the object are devoid of sense and the observer measures all the missing elements that prevent him from evaluating it. When the transmission generated by these things is not ensured, continuous, and moving, the technology is reduced to a handful of objects or isolated texts which, before this, were mixed up in the flow of communication, know-how, habits, and criteria for judging, organizing, and performing various interactions.

Technology in action can therefore be grasped by following and reporting on the action and movement with and by all the intermediaries: tools that are bent and bruised (Linhart 1978), texts that are covered with contradictory and half-finished notes, drawings that never show everything and seem to adhere to tacit conventions, tables of data, and so on.

Overlapping Mediators

Technology that is no longer used, or at least no longer used in a particular place, can be accessed only by a few intermediaries, who are generally not very explicit. Industrial drawing, which is supposed to transmit knowledge about an object, turns out to be of little real use, and can even be misleading. It is thus necessary to re-establish the technology's practical continuity by relocating shared knowledge about the right gestures, the right interpretations and the rules for use as well as the tacit hypotheses (Lavoisy et al. 1998; Lavoisy 2001).

It is notably a question of going back to the languages used to code the technical performance and the objects. From this point of view, technical drawing is an interesting object of study. It corresponds to the construction of a space that can be mathematically expressed and that has partially monopolized the function of referring to the technology itself and its physical objects, by removing them from the immediate wealth of relations that they call into play[18] and then regularly adding new possibilities for expression.[19] Taking into account the language used to mediate involves relocating the social groups, the reasons and interests at stake in their action, the beings they mobilized to provisionally define it (and notably standardization institutions), the final definition and interpretation procedures, and the new practices that are emerging. Technical drawing, like other languages or objects, corresponds to the naturalization and projection on paper of collective practices, which the very properties attached to the object on its own have made invisible.

Although it is important to reconstruct the heat of the action in order to understand the objects, the action can be really grasped only by understanding the inextricable overlappping of the various mediators and the lack of fixation of various bodies, objects, or texts. When the action is part and parcel of a multitude of persons, objects, and texts, technical performance seems only natural. Human control of technology is also contingent on a string of little things. On the other hand, the object suddenly seems to be powerless when deprived of its producers, maintainers, and users. Technology is, intrinsically, as powerless as human beings who are naked and weak without their reference points, objects, and texts. Elements able to impose themselves without any help are few and far between.

Autonomous technical objects, like free human subjects, are divested of their force when they lose the relations that maintain them and which they maintain. Objects and gestures are mixed up. Action programs on matter are action programs on society, and vice versa. As we have seen with design activities, even thought operations and sequences of ideas are spread through situations, objects, and groups. Cognition and imagination belong to sets of things and people; they are not imprisoned in the brain, neither are they confined to a transcendent symbolic system.

Now, all relations to things suppose some kind of organizational structure, translation, and hybridization. Relations to objects are no more immediate than those between humans. They must be built and consolidated. They must be objectivized, by intermingling the constraints and forces of numerous mediators, and re-configured in the process. When beings come into contact with one another, their properties are modified and redistributed. The new mix or the next distribution constitutes a model that forbids, authorizes, or obliges certain associations. It is then no longer possible to do whatever one pleases, since the cards have been dealt and the threads drawn together. Irreversible situations are established, not because the universal laws of nature and those of society deny freedom, but because the specific historical context forbids, allows, or obliges, depending on the web produced.

Concretely, the technical ethnography that we practice tries to follow the operations involved in shaping associated entities. It attempts to follow the same paths used by the objects from their creation to their use and their destruction, and to question the simultaneous formation of technology and its environment, the evaluation systems accompanying them, and the environments and languages that allow them to be qualified. In this way, we avoid separating technical and social universes and transforming them into explicative resources. The actors try to separate realities, to attribute causes to them, to define some of them as the cause of others, and to agree on the general causes and reasons. Technical ethnography must report on the work of these actors.

In-the-Field Approach and Writing

Ethnographic Writing and Technical Writing

Our ethnographic approach is a descendent of debates spawned in the 1970s and the 1980s by young anthropologists who criticized both traditional ethnographic monographs and structuralist analysis.[20] It is now accepted that the reports drawn up are not neutral; they are systematically constructions[21] whose socio-economic, cultural, and political conditions must be exhibited. In the 1980s, anthropologists questioned ethnography as "objective description" and looked into the conventions implicit in traditional ethnography. Notably, they explored the natures of their own discourse (Geertz 1973, 1988).[22] "Reflective" and "critical" essays were then introduced just when, in sociology of the sciences, the question of reflexivity was raised by Bloor (1976), by Latour and Woolgar (1979), and by Lynch (1985).

With respect to ethnography of design activities, it is common for both observers and observed to produce writing. The ethnographer can thus no longer claim to hold real, unquestionable knowledge enabling him to talk about the morals and beliefs of the engineers being studied. The analysis categories and writing conditions of both ethnographers and engineers are all just as questionable. We are thus lead to wonder about the intricate nature of writing and the stakes involved (Stocking 1983) in terms of power and hegemonic relations between our social and technical sciences.

Questioning the role and the place of writing and graphic representation in the universe of technicians sends the ethnographer back toward his own written products: their type, density, readability, circulation, and interpretation by readers. Conversely, thinking about the ethnographical report leads the engineer to question his own technical writing. These different writings call one another into question. How does the ethnographic engineer write differently from his non-ethnographic counterpart, who sits at the same meetings and takes notes? How are the notes of these different people used as resources for understanding, communicating and coordinating? How do their writings model groups, objects, lines of thought, and intentions? Inevitably, the ethnography we practice is reflexive. It assumes that we question our own writing conventions and that we allow the reader to be aware of the constructed nature of our own reports.[23]

One Voice among Others

New generations of anthropologists are more critical of ethnographic writing and its political dimension. For example, Clifford Geertz has been criticized for making an appearance only in his introductions and conclusions and disappearing in the heart of his analyses. This approach is seen to be too conventional, hiding the local and interactive nature of the investigation: it preempts a position of scientific authority and hence re-introduces the former colonial approach that distinguishes between the anthropologist and his indigenous informer.

In our industrial fields, the situation is often different. Here it is no longer a case of a simple colonial or hegemonic relationship. The various relations of superiority, power, and knowledge become much more complex when the identity of the observer-observed pair constantly oscillates between "young observer starting out in research and in the field"-"seasoned technician" and "sociology engineer wreathed in diplomas"-"ordinary technician."

To deal with this issue, anthropologists who are aware of the political dimension in the scientific approach (Clifford 1988) have proposed such alternatives as dialogue-based ethnography and polyphonic ethnography.[24] The ethnographer's voice is thus no longer the only voice to announce a single truth or position itself as the exact reflection of what goes on in the field. At the same time, the subject of the anthropological study is changing. Instead of describing a given culture, the project aims to understand how members of a society build their culture. This is the aim of our project too. We want to understand actors' points of view, how they are structured, and how they interact.

Ethnographers and Technical Informers

There is much debate about dialogue-based and polyphonic ethnography. Some authors accuse these types of ethnography of dealing lightly with the relationship of anthropologists, their field, and informers. Whether the ethnographic narrator sets himself up as a heroic participant observer[25] or as a humble confessor to brilliant informers met in the field, he still remains focused on narrating actions and events and leaves the conditions and framework of the situation in the shadows. Textual reflexivity (incorporating the constructed nature of the text into the text itself) sometimes covers up an absence of theoretical reflection (criticism based on the assumptions, values and categories of its own analysis and that of its colleagues).[26]

Another school to have emerged is that of critical ethnography, which analyzes power relations linked to knowledge and the role of knowledge in rendering things legitimate. Since values and political choices are obviously present in any scientific production, the ethnographer should make a conscious decision about which political policy[27] to adopt. He has to think about the power relations involved in his scientific work. From this point of view, technical ethnography is in a very different position from that of post-colonial anthropology. Our social status as observer-sociologists, for example, is often seen as less prestigious than that of our informers, who are professional people or engineers. These informers talk about our science as a "soft science," as opposed to their own science, which is "hard" or "exact." The power relationship between ethnographers and informers is different from that experienced by anthropologists in the past.

The observer's situation is nevertheless complex; it does not just boil down to a simple power relationship to be analyzed. Sometimes the relationship is more balanced, for example when ethnographers work as engineers within teams of engineers with their own share of responsibility for the success of projects, or when they are paid and assessed by the company they are observing. In these situations, the discourse they use in their scientific work cannot be seen as neutral and authoritarian. They can no more portray their colleagues as different beings from themselves. On the contrary, informers often have the power to elbow us out as undesirable observers, in particular when we are interested in the stakes (political, economic ,and social) of their work,[28] their tools and methods, their ideology, and the values that implicitly guide their choices. Furthermore, some of these informers are the people who order our research, it being for them a resource to be used in action, a resource whose importance and use we do not always measure. More generally, we are committed to and caught up in numerous juxtaposed pairs of actors' games: design/manufacturing, engineering/management, and so on. As observer-participants, our investigations are caught up in and shaken by these relations, which are sometimes difficult to clarify.

We wonder what is the point in writing our reports, reflectively, when we show that they are just one construction among others. Faced with informers who are often dominating with respect to the social sciences, are we not just giving them the means of rejecting our analyses as soft, ideological, subjective, and non-scientific at an even faster rate than before? Similarly, we might ponder on the purpose of producing polyphonic ethnography that hands them the floor when their voices already dominate that of the ethnographer.

Our project is presented as a sort of intrinsic demiurge to technical practices: it involves pinpointing the discrepancy in actors' discourse and putting different types of discourse and points of view into circulation so that they confront one another. Producing various technical work reports means that not only one voice is heard. We add our own voice to the discordant voice of the actors so that it produces collective effects that partially escape us. Our involvement in the field is not only a means of producing academic knowledge; its primary aim is to produce reports and interpretations that can be used by our interlocutors. Within the group observed, the idea is to question the power relations, implicit conventions, and dominated voices (notably those of the technicians, who are not always heard by the decision makers, and those of the non-human elements—graphics, tools, places—that receive very little discussion). If the critical approach to which the researcher is committed in the field sometimes raises problems, this is to be expected in most cases. The reports are supposed to offer another view, which our informers say they need. Our involvement often leads us to provide the actors with elements of assessment, to suggest that other parameters be incorporated, and to validate, with them, a number of hypotheses and analyses and to guide them in their action.

Conclusion

The ethnography of design activities involves looking at our society differently.

Technical objects interfere with the familiar distinctions between the natural and artificial and between the human and the non-human. They are non-human, but they are made by men and women. They are the concrete results of social actions with a physical consistency, strangely human machines or technical monsters. What can we say, then, about humans with standard and repetitive movements, dressed to the nines in technology? The anthropological strangeness of such beings can be taken as a model of anything or anyone: systematically made up of other beings, these beings are collective and hybrid (Callon and Law 1993). Thus, the strangeness switches over to the side of pure beings, converted into mythical figures impossible to find: the human being himself, the symbol, the thing itself, the pure technicality, and so on.

Hence, our conception of the relations between technology and its environment is problematic. In the past, authors were keen to understand a specific type of technology, which could be isolated from its environment. This environment was itself broken down into an "associated environment" (Simondon 1989) and an undefined or indifferent environment. But the objects cannot be isolated; they are linked to other objects by numerous interdependent relations. Taken by itself, the object has meaning only within a network. It is just one element linked to a node in a network from which society cannot be excluded. This socio-technical network is like the underground mycelium, and the objects, texts, or persons are like the mushrooms growing above ground. Or, to use another metaphor, they are the terminals in a vast electronic or telephone network. On their own they have no meaning. As a node of relations, their meaning emerges from the web of numerous links that make them up.

From this point of view, modern technologies have more social threads than old technologies. It is possible to imagine a hammer isolated from the society that produces it, but not a telecommunications network. Not only are they formed through complex socio-technical processes and controversies; in addition, our societies lend them an increasing multitude of properties: physical force, upholding of social relations, social monitoring, moral reminders, intelligence, fidelity, skillfulness, and so on (Latour 1993).

By following the movements of mediation, it is possible to carry out unified reporting on situations ranging from manager specification to technical innovation and social negotiation. This means that there will be no difference in nature, only in composition between a management technique and a machine during operation. Whatever the case, it is a question of implementing a stable and imperative program of action, which is assumed to be rational—that is, thought out, weighed up, negotiated, optimized, translated, and kept by a multitude of intermediaries.

Machine technology, far from boiling down to the materials and rational principles governing it, includes as many principles of management and use as any other management technique. Inversely, management techniques can be broken down into a wide variety of ad hoc instruments, which harden it and give it a specific activity and a specific level of efficiency.

Replacing the relation between object and environment with the web of mediation networks thus leads to a new understanding of technology and of society. By circling above the single sphere of industrial production in order to gradually get to the bottom of our technical societies, studying technology becomes a central issue in sociology, since there are hardly any "social" facts that are not today transformed into sociotechnical realities.

Notes

Introduction

1. The laboratories are (1) CRISTO (Center de Recherche: Innovation Socio-Technique et Organization industrielle), affiliated with the Université Pierre Mendès-France and the Centre national de la recherche scientifique (CNRS); (2) 3S (Sol-Solide-Structure), affiliated with the École Nationale Supérieur d'Hydraulique et de Mécanique de Grenoble, the CNRS, and the Université Joseph Fourrier; and (3) SEED (Socio-Economie Environnement et Développement), affiliated with the Fondation Universitaire Luxembourgeoise (Arlon, Belgium).

Chapter 1

1. This chapter sums up work by Grégoire Pépiot, Jean-François Boujut, Pascal Lécaille, and Bertrand Nicquevert. Grégoire Pépiot is a mechanical engineer studying for a research diploma in industrial engineering under the supervision of Jean-François Boujut (a mechanical engineer) and Dominique Vinck (a sociologist). Bertrand Nicquevert heads a technical office in the Experimental Physics Division of CERN in Geneva.

2. As a part of their studies, young engineers are placed with companies. The idea is that this gives them an opportunity to put what they have learned into practice.

3. A DEA (Diplôme d'Etudes Approfondies) is a one-year postgraduate research diploma. It can be done at the same time as a PFE (Projet de Fin d'Études—a placement project carried out by engineers in their final year of study).

4. The head of the design office does not share this opinion. On the contrary, he says that there are very few procedures involved.

5. In fact, he didn't think it was up to him to define the various wishes of the physicists.

6. The head of the design office says that this is not really a problem since nobody has overall authority over the people involved in the project. Indeed, the way physical research is organized at the end of the 20th century is based on partnership, which means going through a long series of discussions to reach a consensus about what is possible (mechanically, geometrically, and perhaps financially).

7. This harks back to the title of Peter Galison's book *How Experiments End.*

Chapter 2

1. The French term for computer-aided design is *conception assistée par ordinateur,* abbreviated CAO.

2. In French it is SCT (Standardisation et Coordination Technique).

3. The French term for computer-aided design–computer-aided manufacturing is *conception et fabrication assistées par ordinateur,* abbreviated CFAO.

4. Here CAD stands for both computer-aided design and computer-aided drawing. The French acronym, DAO-CAO, stands for *dessin assisté par ordinateur–conception assistée par ordinateur.*

5. Especially with respect to electrical resistance. Green is, after all, a manufacturer of low-voltage to extra-high-voltage electrical equipment.

Chapter 3

1. This reduction was viewed in terms of the treatment of collected waste. However, human intervention was increased earlier in the process: users were asked to do more sorting, and extra handling by waste collectors was necessary.

2. See Vinck and Jeantet 1995.

3. There are two explanations for the fact that no partition is initially provided: (1) Overlapping contracts make the District Council dependent on the material the private collector's supplier can provide. For the price the private collector is willing to pay, the supplier is able to provide only containers with no compartments. (2) A problem arises concerning the manufacturing license of the containers. The Duobac containers are licensed for partitions, but with another supplier. When the private collector's supplier fits containers that were not designed for this purpose with a partition, the licensed supplier sues him. The end result is nevertheless that the private collector, working in collaboration with the District Council, supplies containers fitted with makeshift partitions when nothing in the manufacturing plan or the actual containers produced allowed for the fitting of these partitions. Therefore, iron fixtures have to be manually installed in each container so that the initially unforeseen partitions can be put in position and play their role. The partitions do not, however, completely fill

their role, as liquid from biodegradable elements seeps under the separating wall and contaminates waste that is intended for incineration.

Chapter 4

1. Some clients still have their own bearing design departments. Most of these are engine manufacturers. Bearings are essential to the operation of an engine, and they determine its size.

2. The situation is changing as a result of pressure on costs exerted by clients. This trend has thrown doubt on design office practice at Air Bearings.

3. The design office is home to two categories of actors: structural engineers and CAD operators.

4. In the aeronautics sector, prototypes must be manufactured using the same tools as will be used to produce subsequent production runs. In this way the product and the production system are both assessed. In addition, orders for prototypes are far from negligible in economic terms, as the sales price is calculated in such a way as to cover a large part of the cost of design work.

5. For a detailed description of this process, see Mer 1998.

6. There is no point in trying to decide whether the manner in which a bearing is represented influenced simulation software, or vice-versa. The important thing is to recognize that they are closely linked.

7. It is nevertheless a part of the structural engineers' world, just as the draftsmen are part of the design office. There is a specific room, known as the CAD room, that houses the design workstations. The structural engineers do not often venture in there, except to fetch a printout.

8. I refer here to the two action motivations defined by Weber: value-rational and purposive-rational.

9. The theoretical side was already present, but empirical knowledge played a larger part than it does now.

10. Machine tools require high-precision bearings, which is why they are designed and manufactured by the Air Bearings division.

11. Weight is a crucial factor in aeronautics; a product can be priced higher if it saves weight.

12. These mechanisms use part of the jet engine's power to supply the plane with electricity. In helicopters they drive the tail-rotor blades.

13. Air Bearings is not competitive for all types of bearings. Thus, it concentrates on certain selected types.

14. For a presentation of the various worlds to be found at Air Bearings, see Mer 1998. My aim here is to show that the relations between them feed the dynamics.

Chapter 6

1. In outline, the technology of air bearings and air thrust bearings consists in injecting a film of air a few hundredths of a millimeter thick between stationary and mobile parts. If the air supply is cut off, the mobile and stationary elements may come into contact and be damaged if the relative speed increases.

2. The magnetic bearing consists of a closed magnetic circuit in which the flow is looped. The circuit is identical for the stator part and the rotor part, i.e., it is made up of two concentric magnetized rings, connected by a magnetic bridge of soft iron. Thanks to this system, the bearing is in an axial position. The mobile and stationary parts are separated by an air thrust bearing (a film of air a few hundredths of a millimeter thick).

3. These machines use air bearings that allow high rotation speeds.

4. The law of physics on which this spindle is based is practically the same as that on which the permanent-magnet spindle is based. Radial stiffness is achieved by looping the flow inside a magnetic system. It's the development of the magnetic system that differs. Instead of using two pairs of permanent magnets, the magnetic properties of ferromagnetic materials are employed (in this case, "soft iron"). Only one magnet is then required to create the magnetic flow.

5. The element known as the dumper is in fact a shock absorber fitted between the spindle's body and the stator magnets. On the first spindles, it was developed simply, using rubber pipes, but these were not resistant to solvents. Nowadays, the stator magnets float inside the body of the spindle. Besides absorbing shocks and giving clearance, the dumper allows the automatic centering of magnets in relation to the body of the spindle.

Chapter 7

1. As the anthropologist Jack Goody (1980) has shown, the use of writing, the type of written document, and the techniques used for copying have consequences, which are often indirect and complex, for the distribution of knowledge and power. In sociology of sciences (Latour and Woolgar 1998; Vinck 1992) and in sociology of techniques (Jeantet 1998; Vinck 1999), several authors have considered the extensive variety of graphic inscriptions in order to take into account the activities they are studying. They invite the reader to take a closer look at the writing supports and forms (inscriptions, traces, unchanging and combinable mobiles, intermediary objects) in order to understand the relations between those involved, the networks and the effective organizations. The historian Elisabeth Eisenstein (1983), working on the origin of printing, has shown how it transformed scientific activities, how copying made it easier to circulate documents and observations, and how bringing these together and comparing them turned up contradictions between theories and kicked off renewed investigations by researchers. The development of the printing industry also reorganized the market of written documents: advertising for books, payment of authors, etc.

2. We observed a situation where documents were too complex and incoherent due to use of one type of formalism alone. The objective was to incorporate the environment in installation maintenance management. There was a Works Request form which allowed the user to plan operations. This form, already being used, could have continued to have been used except that the manager wanted a logic diagram. He drew one up, gradually including all the elements resulting from discussions, including planning and actions underway. Finally, the diagram was so complex that it was incomprehensible. The fact that only one formalism was used acted as a kind of snare for those involved. It would have been better for them to explore other types of format.

3. Feedback about various serious incidents shows that the weak point of industrial sites more often relates to documentation management problems than to any distance between the operator and the process due to automatic control. One of the most lethal accidents of the last 20 years occurred at the Piper Alpha platform due to an incomplete maintenance operation documentary management procedure.

Chapter 8

1. A solid model provides a geometric view of a part (shape, proportions). It does not give any indication as to the purpose of this part. On the other hand, it does include elements which make sense to those involved in manufacturing. The latter can rapidly discern if the part is supposed to be made by forging or by molding. In the first case the manufacturer will note, for example, the distribution of material mass, and will identify the sides to be machined. This situation also reveals that the designers from the design department base their reasoning on the production procedure that they consider appropriate, and on their knowledge of the processes to be used. Therefore, from the design stage on, they take into account their knowledge of post-manufacturing procedures. This situation illustrates that the experience and knowledge of the people involved are an integral part of their methods and skills (with regard to the design and interpretation of graphic objects). The designers have learned basic manufacturing rules; therefore, professional manufacturers acknowledge their ability to put forward an initial proposal that takes account of manufacturing constraints. It also follows that this accumulation of experience and sharing of knowledge creates design habits. Designers do not start from square one every time. On the contrary, previously acquired knowledge tends to create a design paradigm, which is adjusted according to the manufacturing procedures familiar to the design department. (Laureillard, Boujut, and Jeantet 1998).

2. The spindle supports the wheel hub and the braking system. It must allow the wheel to be turned (for steering).

3. In mountaineering, a hold is a protuberance which the mountaineer uses to hold on to. It therefore refers both to the characteristics of the object, and to the acts performed with it.

4. The assembly planner is responsible for defining the plans (operating modes) for fitting a part to the body of the system to which it belongs.

5. Machining consists in making holes or eliminating excess metal wherever necessary. It completes the forging process, which provides a shape similar to but larger than the final part.

6. The interface agent therefore organizes meetings between the design department and machining method planners. In this case, the interface agent is the trainee mechanic, who is responsible, on an experimental basis, for improving coordination between the various people involved in the project. This coordinating role is being thought out and conceptualized, a process which is not gone into here. The new agent's relevance, mission, institution, and profile have been called into question (Laureillard et al. 1998).

7. A surface model—a nominal representation of the forged product in the form of areas and curves—enables the smith to design the forging die. This type of representation comes into play when the design process is already quite advanced. As modifications to this model are very difficult and time consuming, it is supposed that the shape of the machine part is definitive. This creates an irreversible situation which is incompatible with a clash of opinions. This confrontation is easier with the solid model, which is simpler to modify.

8. Besides, manufacturers, absorbed in the daily activity and emergencies of the workshop, scarcely attach any importance to the work carried out in the design department, which consists in studying future parts. They would have to be removed from their immediate production environment to give priority, and their full attention, to these studies.

9. This is the subject of much debate within engineering science. Geometric representation is criticized for the implicit way in which it expresses technological data. Some authors have therefore suggested that this information should be made explicit through the use of written labels.

10. A machining allowance is the material removed from the forged part in order to produce, with precision, "functional surfaces."

Chapter 9

1. This experiment was carried out as part of a research partnership between mechanics researchers from the ENSHMG's 3S laboratory (Éric Blanco and Olivier Garro) and cognitive psychology researchers from the Communications Research Group at the University of Nancy (Christian Brassac and Nicolas Grégori). This research operation was conducted within the framework of the SPI-SHS (Engineering Sciences–Social and Human Sciences) program set up by the CNRS and focusing on production systems. For further results of this research, see Blanco et al. 1997 and Grégori et al. 1998.

2. Twenty seconds of silence precede the appearance of object 4. Twelve seconds of silence follow the appearance of object 8.

3. "A cognitive artifact is an artificial tool designed to conserve, display and process information with a view to fulfilling a representational role." (Norman 1993, p. 18)

4. Dodier suggests that sociological pragmatism may develop on the basis of the conventional support. A conventional support is a set of resources that favor the creation of a collective point of view and therefore the development of coordinated action. It has its roots in people and in external objects: objects, references, etc. He defines three levels which correspond to three coordination modes: (1) the universal level (this includes coordination models based on the skills common to all human beings, (2) the cultural level (this includes local forms of coordination in communities separated by space and time), and (3) the pragmatic level (this involves conventions that result from the endless adjustments that people make to one another in the course of concrete action). Dodier's basic theory is that action is internally complex, in other words that several coordination modes come together in the course of action. Over time, these coordination modes tie in with one another and form three different categories: simultaneity, succession, and confrontation.

5. When we say that a conventional aid or reference is "solid," we mean that it is acknowledged by everyone involved and that it provides a basis for their actions.

Epilogue

1. In this respect our project differs from that of scientific sociologists in the 1970s who concentrated on establishing a causal relation between a social condition and cognitive content and who therefore explained technical terms and objects according to social categories regardless of the actual methods employed by people. (See Vinck 1995.) Ethnomethodologists strongly contested this approach (Lynch 1985). In their opinion, actions could not be explained by the concept of social groups and their interests, because these factors do not exist before the action occurs. Thus, it was advisable to observe and analyze practices in detail. In fact, until the 1980s very few scientific sociologists paid any attention to specific laboratory practices. Only authors such as Fleck, Polanyi, and Ravetz had studied experimental tools and apparatus, technicians, their know-how, and their tacit know-how. Ravetz believed that scholarly excellence required daily practice and could not be achieved through formal principles. "Although tools are only accessories to the progress of scientific knowledge," he wrote, "their influence on work directions is considerable and often decisive." (Ravetz 1972, p. 89)

2. I refer here to previous work by Elisabeth Eisenstein on the history of printing (1983), by Jack Goody on writing (1980), and by Bruno Latour on scientific writing stratagems (1979).

3. Three ethnographic perspectives are possible here: (1) describe the behavior and words through which human beings create a collective meaning; (2) recreate the symbolic system or grammar which give meaning to behavior; (3) turn out several descriptions until the whole society being studied has been accounted for.

4. According to Lynch (1985), scientific facts cannot be separated from the action that produced them. They cannot be explained by general causes, whether the latter be invisible, technical, or social. It is therefore advisable to describe practices without going any further.

5. This perspective corresponds to an earlier type of anthropology that concentrated on the differences between human societies. From this perspective, the world is made up of peculiarities and strangeness that must clarified. It was usually concerned with other "exotic, primitive or uncivilized" societies, as well as our "modern" ones. Nevertheless, the study of these societies and their differences provided an opportunity to think about Western society.

6. The earliest anthropological theories suggested that societies all evolved in the same way, going from primitive to civilized. Studying exotic societies was tantamount to examining one's own past.

7. In this case, a report isolates a culture, a village or a community from the rest of the world, and it presupposes that the traditional and Western worlds are cut off from one another. The anthropologist is supposed to look at things from a scientific, objective point of view only: he is a hero of science who descends into the field, collects data, and goes home to write up the truth about this society in its various dimensions. It is supposed that he paints a true picture of the society: the aim is to describe societies as they are. Everything is placed in a unified, coherent framework, within which each element fulfills its role. Each dimension is a part of an organic whole.

8. See, e.g., chapters 1, 2, 5, and 8.

9. Traditionally, these dimensions were ecological, economic, social, political, cosmological, aesthetic, and religious.

10. Moreover, in these monographs Thill and Traweek aim less to describe the scientific world such as it is than to question themselves on their own actions. Thill attempts to discover the significance of his own activities as a physicist, with a view to revealing the unsaid and the praxis of scientific theories. In the same way, Traweek's primary aim is to study the conditions and relationships underlying the development of her own interpretation.

11. The structuralist anthropology developed by Lévi-Strauss differs from the rest in that it refuses to take each culture separately, and to consider it as a unique and functionally integrated whole. On the contrary, it compares different cultures, without trying to situate them on a single, evolutionary scale. Therefore, Lévi-Strauss detects and interprets the codes and structures underlying each culture by studying the myths, rites, relationship patterns and systems of distinction that govern them. Structuralism describes the cultural grammar of societies and

explains it as a system of meanings. Thus, studying a society no longer consists in describing it in its functional totality, but in analyzing its underlying structure. However, the structural approach ignores social action, change and history. Society is determined, via thought, by an underlying symbolic structure that only structural formalism can reveal.

12. The weight of a satellite, for example, is now less than the weight of the technical documentation related to it.

13. Those involved in developing the functional dimensioning and tolerancing software for designers (chapter 2) try to bring the software to life through a large number of texts (user guide, diagrams, diagram design methodology), objects (an example of a technical part illustrated by the diagrams), and human beings (trainers, designers-controllers, members of the company's Standardization, Coordination, and Technical department, computer engineers).

14. This analysis is based on Hennion's (1993) work on music. The eminently ephemeral product enshrined in the notion of musical performance has a lot of similarities with the technical performance we are speaking of.

15. A designer explained it to one of us this way: "The engineer designs a ball bearing and translates his or her design into a list of data. The draftsperson takes up this list and adds his or her little bit to a drawing. The production engineering person then gets hold of the drawing and transforms it into a range which looks less and less like the original bearing design. And yet the bearing works."

16. It is also easy to imagine the importance of these mediators when we think about the difficult task of technology museums who strive to produce modern demonstrations of past technologies in the absence of the society of that period. For such exhibitions there is a struggle between, on the one hand, getting the right authenticity to satisfy the experts and collectors and, on the other hand, setting up an original event to urge visitors to take a detour via the past.

17. Consider, for example, the way a mechanical engineering colleague expresses his feelings about a technical part and the power emanating from the quality of its intrinsic fiber orientation.

18. As shown in the history of the progressive separation of artistic drawing from technical drawing.

19. As can be seen with the change from geometrical dimensioning to functional dimensioning.

20. Dissatisfaction was mainly felt by women and anthropologists from the Third World. Two major problems were then underlined: reports drawn up in the colonial and post-colonial context were politically problematic, and bias linked to the sex of the anthropologist seemed to have been greatly underestimated.

21. One indication of their constructed nature is the lack of completeness of the monograph. They very rarely have chapters about relations between the culture

studied and that of the ethnographer, as if the culture studied were completely isolated from the rest of the world, as if the presence of the ethnographer did not change anything, as if the conceptualization of a village as a culture did not reflect just as much the Western anthropologist's culture as that of the "savage." The work of the anthropologist and his way of bringing the raw facts together are missing from these reports. The author is hidden. Although the result of his analysis reflects as much his point of view (and so that of his Western scientific discipline) as the empirical reality that he encounters, the report itself is passed over in silence.

22. Similarly, Geertz (1973) proposes a new perspective for ethnography. Unlike the structuralists, he focuses on rites rather than on myths, on social action rather than on kinship systems. He swaps Lévi-Strauss's linguistic formalism for the anti-formalism of literature. He thus intends to *produce interpretations* of the cultures studied. The ethnographer is a reader of texts who can only understand things by interpreting them or by reconstructing them as he sees fit. Rites and actions are thus considered as texts that can be read in different ways: the reading can be political, economic, psychological, sociological, and aesthetic. The interpretation can, moreover, be reviewed. He proposes to draw up detailed descriptions through which the author's successive interpretations can be built. With each new reading, the author mobilizes part of the data produced and gathered in the field.

23. Another strategy, adopted notably by Woolgar (1988), was to produce reflective essays by exploring new textual forms and introducing doubt and uncertainty concerning what is reported and the way it is reported into the text.

24. Dialogue-based ethnography involves recording dialogues between the ethnographer and his informers and exhibiting his field notes. Thus, by revealing the findings of the field survey, the political and epistemological questions relating to the construction of the analysis can be raised. Polyphonic ethnography, which is apparently more radical, does not see the ethnographer's voice as the only one. The anthropologist's report can thus be contradicted and relativized by transcribing the discourse of his informers, or via a postscript or other right of reply granted to the observed. The anthropologist's authority is hence relativized and referred to the fact that he himself has a specific background: Western or indigenous with a Western-style education for the cultural anthropologist.

25. Some ethnographers have experienced the effects of this risk when reporting our observations. Insofar as they may themselves be engineers, actors, or project initiators, they partially identify themselves with the role they are led to play. During seminars, when talking about the company or the project team working in the field, they all use the word 'we'.

26. The absence of investigation production conditions is extensively criticized by feminist and Marxist anthropologists. Moreover, these anthropologists use ethnography as an instrument for denouncing and criticizing the systems of domination between nations, races, social classes or sexes. Similarly, postmodern anthropologists use ethnography to show that worlds and cultures are made up

of controversial codes and meanings. From this point of view, informers and ethnographers do not necessarily share the same projects and have relations based on power. The different languages used by them cannot be separated from the power stakes involved. Because during their construction of "others" and their culture the relationship is always one of domination, critical anthropologists demand that a conscious theoretical framework for the political nature of this relationship be developed at the same time.

27. The critical ethnographer is torn between two strains of thought: feminist and anti-colonialist. Positivist discourse in anthropology is historically linked to colonialism.

28. Compare chapter 6, where Éric Blanco looks into the interactions between the world of the design office where he is responsible for a design assignment and the world on which the company strategy, with its strict access rules, is based.

Bibliography

Akrich, Madeleine. 1992a. Beyond social construction of technology: The shaping of people and things in the innovation process. In *New Technology at the Outset*, ed. M. Dierkes and U. Hoffman. Westview.

Akrich, Madeleine. 1992b. The de-scription of technical artifacts. In *Shaping Technology/Building Society*, ed. W. Bijker and J. Law. MIT Press.

Akrich, Madeleine. 1995. User's representations: Practices, methods and sociology. In *Managing Technology in Society*, ed. A. Rip et al. Pinter.

Bagla-Gökalp, Lusin. 1996. Le chercheur et son instrument. *Revue Française de Sociologie* 37, no. 4: 537–566.

Becker, Howard S. 1982. *Art Worlds*. University of California Press.

Béguin, Pascal. 1997. Les technologies de l'information: Dématérialisation ou nouvelles formes de matérialité des objets de travail. In *Intégration du savoir-faire*, ed. J.-M. Fouet. Hermès.

Bessy, Christian, and Francis Chateauraynaud. 1992. Le savoir-prendre. Enquête sur l'estimation des objets. *Techniques et culture* 20: 105–134.

Bessy, Christian, and Francis Chateauraynaud. 1993. Les ressorts de l'expertise, Epreuves d'authenticité et engagement des corps. In *Raisons pratiques 4, Les objets dans l'action*, ed. B. Conein et al. EHESS.

Bessy, Christian, and Francis Chateauraynaud. 1995. *Experts et faussaires. Pour une sociologie de la perception*. Métailié.

Bijker, Wiebe. 1995. *Of Bicycles, Bakelites and Bulbs: Toward a Theory of Sociotechnical Change*. MIT Press.

Blanco, Éric, Olivier Garro, Daniel Brissaud, and Alain Jeantet. 1996. Intermediary object in the context of distributed design. Paper presented at Computational Engineering in Systems Applications, IEEE-SMC, Lille.

Blanco, Éric, Olivier Garro, and Alain Jeantet. 1997. Conception distribuée approche expérimentale. Paper presented at congres international franco-quebecois de génie industriel, Albi.

Blanco, Éric, Olivier Garro, and Alain Jeantet. 1998. La constuction conjointe du problème et des solutions en ingénierie simultanée. Paper presented at Journée Primeca Lyon Concurrent Engineering.

Blanco, Éric, Alain Jeantet, and Jean-François Boujut. 1996. Copest de la construction à l'usage et vice versa. *Revue Sciences et Techniques de la Conception* 5, no. 2: 93–118.

Bloor, David. 1976. *Knowledge and Social Imagery*. Routledge & Kegan Paul.

Boltanski, Luc, and Laurent Thevenot. 1987. *Les économies de la grandeur*. Presses Universitaires de France.

Boujut, Jean-François, Serge Tichkiewitch, and Éric Blanco. 1997. Integration of downstream actors in the design process using a dedicated expert CAD tool for forged parts. *Concurrent Engineering Research and Applications* 5, no. 4: 327–337.

Bucciarelli, Louis L. 1994. *Designing Engineers*. MIT Press.

Button, Graham. 2002. The ethnographic tradition and design. *Design Studies* 21: 319–332.

Callon, Michel. 1988. *La science et ses réseaux*. Découverte.

Callon, Michel, and John Law. 1993. Agency and the hybrid collective. Paper presented at Theory and Method Non-Human Agency, University of Surrey.

Campinos, M., and C. Marquette. 1999. Une rationalisation sans norme organisationnelle: la certification ISO 9000. *Sciences de la Société* 46, February: 83–98.

Chapel, Vincent. 1997. La croissance par l'innovation intensive: de la dynamique d'apprentissage à la révélation d'un modèle industriel; le cas TEFAL. Ph.D. dissertation, Ecole des Mines de Paris.

Clifford, J. 1988. *The Predicament of Culture*. Harvard University Press.

Constant, Edward W. 1980. *The Origins of the Turbojet Revolution*. Johns Hopkins University Press.

Dodier, Nicolas. 1992. Les appuis conventionnels de l'action: Eléments de pragmatique sociologique. *Réseaux* 62, November-December: 65–85.

Dodier, Nicolas. 1995. *Les hommes et les machines*. Métaillé.

Downey, Gary. 1992. CAD/CAM saves the nation? Toward an anthropology of technology. In *Knowledge and Society*, ed. D. Hess and L. Layne. JAI.

Eisenstein, Elizabeth L. 1993. *The Printing Revolution in Early Modern Europe*. Cambridge University Press.

Forsythe, Diana. 1993. Engineering knowledge: The construction of knowledge in artificial intelligence. *Social Studies of Sciences* 23: 445–447.

Freyssenet, Michel. 1992. Processus et formes sociales d'automatisation. Le paradigme technologique *Sociologie du Travail* 4: 469–496.

Galison, Peter. 1987. *How Experiments End*. University of Chicago Press.

Garfinkel, Harold. 1967. *Studies in Ethnomethodology*. Prentice-Hall.

Garro, Olivier, Daniel Brissaud, and Éric, Blanco. 1998. Design criteria. In proceedings of Ninth Symposium on Information Control in Manufacturing, Nancy-Metz.

Geertz, Clifford. 1973. *The Interpretation of Cultures*. Basic Books.

Geertz, Clifford. 1988. *Works and Lives: The Anthropologist as Author*. Stanford University Press.

Goody, Jack. 1980. *La raison graphique*. Minuit.

Gorges, Irmela. 1996. The impact of society on CAD research in the USA, France and Germany, 1955 through 1985. In *The Role of Design in the Social Shaping of Technology*, ed. J. Perrin and D. Vinck. European Commission Directorate-General, Research and Development.

Gras, Alain, and C. Moricot, eds. 1992. *Technologies du quotidien. La complainte du progrès*. Autrement.

Grégori, Nicolas, Éric Blanco, Christian Brassac, and Olivier Garro. 1998. Analyse de la distribution en conception par la dynamique des objets intermédiaires. In *Les objets en conception*, ed. B. Trousse and K. Zreik. Hermès.

Grosjean, Michèle, and Michèle Lacoste. 1998. L'oral et l'écrit dans les communications de travail ou les illusions du 'tout écrit.' *Sociologie du travail* 4: 439–461.

Hatchuel, Armand. 1994. Apprentissages collectifs et activités de conception. *Revue Française de Gestion*, June-July: 109–115.

Henderson, Kathryn. 1991. Flexible sketches and inflexible data bases: Visual communication, conscription devices and boundary objects in design engineering. *Science, Technology, & Human Values* 16, no. 4: 448–473.

Hennion, Antoine. 1993. *La passion musicale. Une sociologie de la médiation*. Métaillé.

Hess, David. 1992. The new ethnography and the anthropology of science and technology. In *Knowledge and Society*, ed. D. Hess and L. Layne. JAI.

Hutchins, Edwin. 1995. *Cognition in the Wild*. MIT Press.

Jeantet, Alain. 1998. Les objets intermédiaires dans les processus de conception des produits. *Sociologie du travail* 3: 291–316.

Karpik, Lucien. 1972. Les politiques et les logiques d'action de la grande entreprise industrielle. *Sociologie du travail* 1: 71–95.

Kaufmann, Jean-Claude. 1992. *La trame conjugale. Analyse du couple par son linge.* Nathan.

Kuhn, Thomas S. 1968. *The Structure of Scientific Revolutions.* University of Chicago Press.

Latour, Bruno 1991 Technology is society made durable. In *A Sociology of Monsters,* ed. J. Law. Routledge.

Latour, Bruno. 1993. *We Have Never Been Modern.* Harvard University Press.

Latour, Bruno. 1994. Une sociologie sans objet? Remarques sur l'interobjectivité. *Sociologie du travail* 36, no. 4: 587–607.

Latour, Bruno, and Pierre Lemonier, eds. 1994. *De la préhistoire aux missiles balistiques. L'intelligence sociale des techniques.* Découverte.

Latour, Bruno, and Schirley Strum. 1991. Towards a common genealogy for humans and their artefacts. Mimeographed abstract.

Latour, Bruno, and Steve Woolgar. 1988. *Laboratory Life: The Construction of Scientific Facts.* Princeton University Press. (First edition: 1979.)

Laureillard, Pascal, Jean-François Boujut, and Alain Jeantet. 1998. Conception intégrée et entités de coopération. In *Les objets en conception,* ed. B. Trousse and K. Zreik. Hermès.

Lavoisy, Olivier. 2001. Coordination by design: Key roles of technical graphics in industrial France. Paper presented at Design History Society Conference, Victoria and Albert Museum/Royal College of Art, London.

Lavoisy, Olivier, Dominique Vinck, Laurent Gauthier, and Patrick Lacheau. 1998. Designing a CAD-tool: A participative observation of a distributed work. In proceedings of Ninth Symposium on Information Control in Manufacturing, Nancy-Metz.

Lemonier, Pierre, ed. 1993. *Technological Choices: Transformation in Material Cultures since the Neolithic.* Routledge.

Linhart, Robert. 1978. *L'établi.* Minuit.

Lynch, Michael. 1985. Discipline and the material form of images: Analysis of scientific visibility. *Social Studies of Sciences* 15: 37–66.

Mer, Stéphane. 1998. Les mondes et les outils de la conception. Pour une approche sociotechnique de la conception de produit. Ph.D. dissertation, Institut National Polytechnique de Grenoble.

Mer, Stephane, Alain Jeantet, and Serge Tichkiewitch. 1995. Les objets intermédiaires de la conception. In *Le communicationnel pour concevoir,* ed. J. Caelen and K. Zreik. Editions Europia Productions.

Midler, Christophe. 1994. *L'auto qui n'existait pas, Management des projets et transformations de l'entreprise.* InterEditions.

Moisdon, Jean-Claude, ed. 1997. *Du mode d'existence des outils de gestion. Les instruments de gestion à l'épreuve de l'organisation.* Seli Arslan.

Norman, Donald. 1993. Les artefacts cognitifs. In *Raisons pratiques 4, Les objets dans l'action,* ed. B. Conein et al. EHESS.

Orr, Julian E. 1996. *Talking about Machines An Ethnography of a Modern Job.* Cornell University Press.

Pahl, G., and W. Beltz. 1996. *Engineering Design: A Systematic Approach,* second edition. Springer.

Peterson, Roger, Guy Mountfort, P. A. D. Hollom, and Paul Géroudet. 1994. *Guide des oiseaux de France et d'Europe.* Delachaux et Niestlé. (First edition published in 1954.)

Piore, Michel, R. Lester, F. Kofman, and K. Malek. 1997. L'organisation du développement des produits. In *Les limites de la rationalité,* volume 2, ed. B. Reynaud. Découverte.

Poitou, Jean-Pierre. 1984. Dessin technique et division du travail. *Culture technique, Les ingénieurs* 12: 196–207.

Poitou, Jean-Pierre. 1989. *30 ans de CAO en France.* Hermès.

Poitou, Jean-Pierre. 1997. La gestion collective des connaissances et la mémoire individuelle. In *Intégration du savoir-faire,* ed. J.-M. Fouet. Hermès.

Prost, Antoine. 1996. *Douze leçons sur l'histoire.* Seuil.

Ravetz, Jerome. 1972. *Scientific Knowledge and its Social Problems.* Clarendon.

Scardigli, Victor. 1992. *Le sens de la technique.* Presses Universitaires de France.

Segrestin, Denis. 1996. Sur la 'Traduction' des normes de gestion de la qualité dans l'entreprise: Premières observations et analyses. Paper presented at Congrès International du Génie Industriel, Grenoble.

Segrestin, Denis. 1997. *Sociologie de l'entreprise.* Armand Colin.

Simondon, Georges. 1989. *Du mode d'existence des objets techniques.* Aubier. (First edition published in 1958.)

Star, Susan Leigh. 1989. The structure of ill-structured solutions: Heterogeneous problem-solving, boundary objects and distributed artificial intelligence. In *Distributed Artificial Intelligence,* volume 2, ed. M. Huhns and L. Gasser. Morgan Kaufman.

Stocking, George W., Jr. 1983. *Observers Observed: Essays on Ethnographic Fieldwork.* University of Wisconsin Press.

Thill, Georges. 1973. *La fête scientifique.* Aubier Montaigne.

Tornatore, Jean-Luc. 1991. Etre ouvrier de la Navale à Marseille. *Terrain* 16, March: 88–105.

Traweek, Sharon. 1988. *Beamtimes and Lifetimes: The World of High Energy Physicists.* Harvard University Press.

Ullman D., S. Wood, and D. Craig. 1990. The importance of drawing in the mechanical design process. *Computer and graphics* 14, no. 2: 263–274.

Vinck, Dominique. 1992. *Du laboratoire aux réseaux. Le travail scientifique en mutation.* Office des Publications de la CCE.

Vinck, Dominique. 1994. Etre objet parmi les autres. *Turbulences* 1: 16–21.

Vinck, Dominique. 1995. *Sociologie des sciences.* Armand Colin.

Vinck, Dominique. 1999. Les objets intermédiaires dans les réseaux de coopération scientifique. Contribution à la prise en compte des objets dans les dynamiques sociales. *Revue Française de Sociologie* 40, no. 2: 385–414.

Vinck, Dominique, and Alain Jeantet. 1995. Mediating and commissioning objects in the sociotechnical process of product design: A conceptual approach. In *Management and New Technology: Design, Networks and Strategies,* ed. D. MacLean et al. European Commission Directorate-General, Research and Development.

Vinck, Dominique, and Pascal Laureillard. 1995. Coordination par les objets dans les processus de conception. In *Représenter, coordonner, attribuer,* Actes des Journées CSI-Ecole des Mines, Paris.

Weil, Benoît. 1999. Conception collective, coordination et savoirs. Les rationalisations de la conception automobile. Ph.D. dissertation, Ecole des Mines de Paris.

Wittgenstein, Ludwig. 1958. *Philosophical Investigations.* Blackwell.

Woolgar, Steve. 1985. Why not a sociology of machines? The case of sociology and artificial intelligence. *Sociology* 19: 557–572.

Woolgar, Steve, ed. 1988. *Knowledge and Reflexivity: New Frontiers in the Sociology of Knowledge.* Sage.

Index

Printed in the United States
by Baker & Taylor Publisher Services